ENZYMIC CATALYSIS

ENZYMIC CATALYSIS

ENZYMIC CATALYSIS

PROCEEDINGS OF
A ROYAL SOCIETY DISCUSSION MEETING
HELD ON 5 AND 6 DECEMBER 1990

ORGANIZED AND EDITED BY
A. R. FERSHT AND D. GANI

LONDON
THE ROYAL SOCIETY
1991

Printed in Great Britain for the Royal Society
by the
University Press, Cambridge

First published in *Philosophical Transactions of the Royal Society of London,*

series B, volume 332 (no. 1263), pages 105–184

The text paper used in this publication is alkaline sized with a coating which is predominantly calcium carbonate. The resultant surface pH is in excess of 7.5, which gives maximum practical permanence.

British Library Cataloguing in Publication Data

A CIP catalogue record for this book is available from the British Library.

ISBN 0 85403 439 0

Published by the Royal Society
6 Carlton House Terrace, London SW1Y 5AG

CONTENTS

Polyketide synthase complexes: their structure and function in antibiotic biosynthesis

JOHN A. ROBINSON

Organisch-Chemisches Institut, Universität Zürich, CH 8057 Zürich, Switzerland

SUMMARY

This paper gives an overview of existing knowledge concerning the structure and deduced functions of polyketide synthases active in antibiotic-producing streptomycetes. Using monensin A as an example of a structurally complex polyketide metabolite, the problem of understanding how individual strains of microorganism are 'programmed' to produce a given polyketide metabolite is first outlined. The question then arises, how is the programming of polyketide assembly related to the structural organization of individual polyketide synthase complexes at the biochemical and genetic levels? Experimental results that help to illuminate these relations are described, in particular, those giving information about the structures and deduced functions of polyketide synthases involved in aromatic polyketide biosynthesis (actinorhodin, granaticin, tetracenomycin, *whiE* spore pigment and an *act* homologous region from the monensin-producing organism), as well as the macrolide polyketide synthase active in the biosynthesis of 6-deoxyerythronolide A.

1. INTRODUCTION

The polyketide hypothesis enunciated by Birch stands out as an incisive contribution to the field of polyketide biogenesis (Birch & Donovan 1953; Birch 1957). It provided a valuable correlation between the structures of a large number of (largely) aromatic natural products and their probable modes of biosynthesis through the head-to-tail linkage of acetate units. The hypothesis was firmly grounded in mechanistic chemistry, and showed how a carbon chain of β-ketone groups, retained from the successive condensation of acetate units (instead of being serially removed as in fatty acid biosynthesis), might undergo well-precedented laboratory reactions, including aldol condensation, C-acylation, reduction, dehydration, methylation and oxidation, and thereby give rise to the large family of known polyketide metabolites. The biogenetic correlation was most apparent for certain plant phenolic compounds that have oxygen substituents attached β- to each other or to positions of ring closure, as a direct result of the β-positioning of ketone and methylene groups in the original chain. Birch also suggested that coenzyme-A esters other than acetyl-CoA might initiate chain assembly, while 'propionate' units (methylmalonyl-CoA) could be incorporated in place of malonyl-CoA during chain extension (Birch 1967).

Only a few years before, the first macrolide antibiotic had been reported by Brockmann & Henkel (1951), and the relevance of Birch's polyketide hypothesis to macrolide biosynthesis was quickly recognized. By 1957 at least six different macrolides had been discovered, and Woodward suggested that the macrolide rings could also arise by the stepwise condensation of acetate (or related) building blocks, with the oxygen atom of individual β-ketone groups retained as hydroxyl groups in the macrolide backbone (Woodward 1957). Since that time close to two hundred different macrolides have been isolated, largely from microorganisms of the genus *Streptomyces*, and many other important classes of antibiotics are now also known to be of polyketide origin. It is a remarkable aspect of natural products chemistry today that new, interesting, and therapeutically important polyketide metabolites continue to be discovered in Streptomycete screening programmes that are conducted largely within the pharmaceutical and agrochemical industries (e.g. avermectins, FK506, etc.).

The main focus of this paper, however, is not the chemistry of polyketides per se, but rather the enzymology of polyketide biogenesis as revealed over the past few years through the application of molecular genetic techniques. Although a good understanding of the central underlying chemical concepts in polyketide assembly had been laid by Birch's polyketide hypothesis, very little was known until recently about the structure of the polyketide synthases (PKSs) that catalyse carbon chain assembly, nor how their structure might be related to the 'programming' that directs individual strains of microorganism to produce a unique polyketide antibiotic. This problem of understanding PKS programming is discussed below with reference to the polyether ionophore antibiotic monensin A (an antibiotic of interest in the authors laboratory).

2. THE BIOSYNTHESIS OF MONENSIN; AND POLYKETIDE SYNTHASE PROGRAMMING

Monensin A is essentially a long highly functionalized fatty acid incorporating ether rings to enforce a

Phil. Trans. R. Soc. Lond. B (1991) **332**, 107–114
Printed in Great Britain

[1]

107

8-2

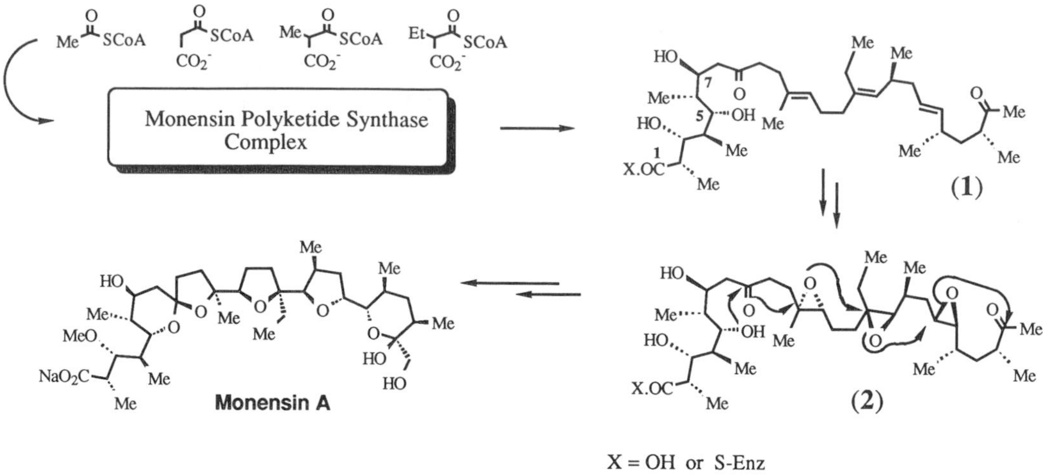

Figure 1. A hypothetical pathway to monensin A.

folded backbone conformation for binding Na^+ or K^+ cations. Carbon-13 labelling experiments revealed how five acetate units, seven propionates and one butyrate are linked in a head-to-tail (i.e. C(1) to C(2)) fashion during formation of the carbon backbone. The 3MeO methyl group is derived in addition from methionine, although the timing of methylation is uncertain. Similar experiments using ^{13}C- and ^{18}O-doubly labelled acetate and propionate, as well as fermentations under $^{18}O_2$, later defined the origins of all the oxygen atoms in monensin A (Cane *et al.* 1982; Ajaz *et al.* 1987). This information provided support for an attractive mechanism to account for ether ring formation, shown in figure 1, involving the triene (1) and triepoxide (2) as intermediates (Cane *et al.* 1983). The fact that each of the β-placed oxygen atoms from C1 through to C9 is incorporated intact shows that the absolute configuration of the C(3), C(5), and C(7) centres is established directly through stereodivergent reductions of β-ketone intermediates, rather than by additional dehydration and rehydration steps. Because C(3) is *R*, and C(5) and C(7) are *S*, this would seem to require at least two different ketoreductases having opposite stereospecificities. Similar results have been obtained in studies of macrolide biosynthesis, strengthening the idea of a common mode of stereocontrol during the assembly of these two classes of polyketide antibiotics (for a review see Robinson (1988)). In the biosynthesis of macrolides, and presumably also polyethers, it is after completion of carbon chain assembly that the backbone is oxygenated, or glycosylated, steps that amplify the scope for structural diversity within each class of antibiotic.

The triene (1) is the presumed end product of a series of condensation, reduction and dehydration events, utilizing substrates derived from acetate, propionate and butyrate, catalysed by the monensin PKS multienzyme complex. An important question concerns the order of these various events during chain assembly. In macrolide biosynthesis, evidence has accumulated recently in support of the so-called processive stategy of backbone assembly (Yue *et al.* 1987; Cane & Yang 1987). In this strategy, the β-ketoacyl thioester formed by decarboxylative-condensation of an acyl chain with a malonyl, methylmalonyl, or ethylmalonyl extender

unit, is modified (where necessary) before the next building block is incorporated. The modification can involve: reduction only to a β-hydroxythioester; reduction and dehydration to give an olefin; or reduction, dehydration and further reduction to give a fully saturated chain (see figure 2). However, the same modification need not occur during each cycle of chain extension. The evidence supporting this model of macrolide assembly includes the incorporation of putative chain elongation intermediates into tylactone (Yue *et al.* 1987), erythromycin B (Cane & Yang 1987) and nargenicin (Cane & Ott 1988), as well as the isolation of compounds considered to be chain elongation intermediates in the biosynthesis of protomycinolide IV (Takano *et al.* 1989). It is very likely that a processive strategy of chain assembly also operates during polyether biosynthesis, although successful incorporation experiments in support of this have not yet been described.

Chemical, biochemical and more recently molecular genetic studies have served to emphasize a fundamental similarity between fatty acid synthases (FASs) and PKSs, both in terms of their mechanisms of action and molecular organization. However, the synthesis of a polyketide must be more highly programmed than that of a fatty acid. For example, during assembly of the putative monensin triene intermediate (1), the PKS must: select an acetate starter unit (presumably acetyl-CoA) and then 12 further building blocks in a specific order from activated acetyl, propionyl and butyryl residues; incorporate the correct chemical functionality during each chain extension step, leaving a ketone, hydroxy, enoyl or saturated alkyl group; and establish the correct relative and absolute configuration at each new chiral centre and double bond. Leaving aside the question of stereochemistry, this programming must be translated into a series of choices regarding the type of extender unit, and the type of chemistry after each subunit addition, as shown in figure 2. Related schemes could be compiled in a similar way to arrive at other polyether intermediates, or macrolide rings. An important objective of current research is to understand how the programming, which allows the monensin PKS to assemble just one out of an enormous number of theoretically possible reduced polyketide chains, is

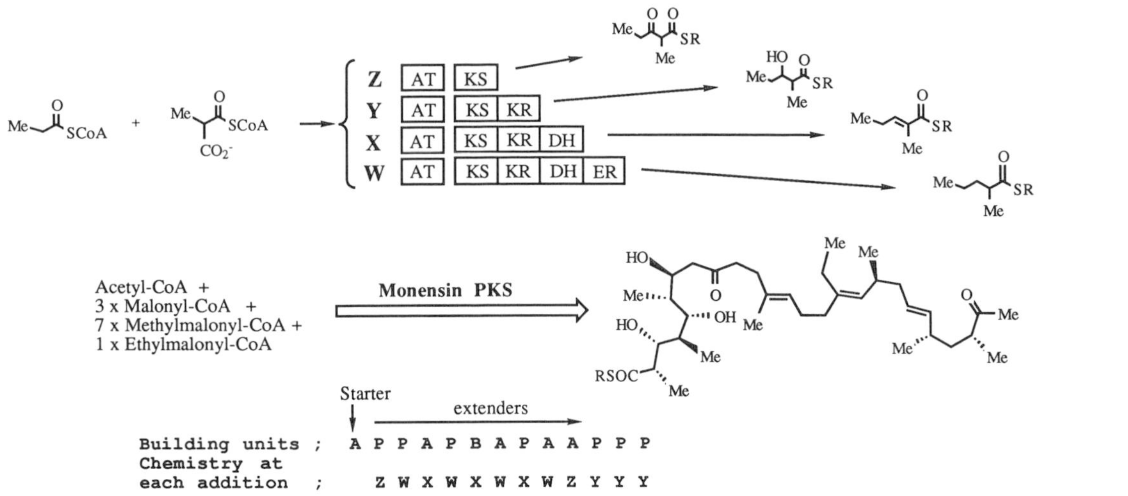

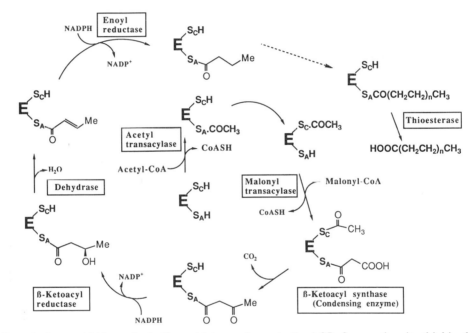

Figure 2. The monensin PKS must select the desired building unit (A = acetate, P = propionate, B = butyrate) and the correct chain extension chemistry (AT = acyl transferase, KS = β-ketoacyl synthase, KR = β-ketoacylreductase, DH = dehydrase, ER = enoyl reductase) at each round of chain elongation.

Figure 3. Steps in fatty acid biosynthesis (S_A = thiol on cofactor in the ACP; S_C = active site thiol in the condensing enzyme).

related to the structural organization of the polyketide synthase complex, at the biochemical and genetic levels.

3. THE STRUCTURE OF TYPE-I AND TYPE-II FATTY ACID SYNTHASES

In discussing the enzymology of PKS function it is helpful to draw upon comparisons to fatty acid synthases, where the relations between structure and function are becoming more clearly understood. The component activities of the type-II fatty acid synthases from plants and bacteria can be readily separated into discrete globular proteins, which together catalyse the consecutive steps of the fatty acid synthase cycle (see figure 3) (Fulco 1983). There is no evidence so far that

these proteins form an aggregate within cells, and this is certainly not necessary to observe full fatty acid synthase activity *in vitro*. The following proteins compose the *E. coli* FAS responsible for palmitic acid biosynthesis (Volpe & Vagelos 1973): (i) the acyl carrier protein (ACP), a relatively small protein of around 80 amino acids, containing the phosphopantetheine cofactor. It plays a special role as it must bind the substrate at most steps of the cycle, and it must interact with all the component enzymes in the fatty acid synthase complex; (ii) and (iii) acetyl and malonyl transferases that load the substrates acetyl-CoA and malonyl-CoA onto ACPs, each via a mechanism involving discrete O-acyl(Ser)-enzyme intermediates: (iv) condensing enzymes (β-ketoacyl synthases), of which three have been isolated from *E.*

coli, that differ in their substrate specificities. The substrates for the synthase-I and synthase-II, both homodimeric proteins, are an acyl-ACP and malonyl-ACP, whereas the synthase-III condenses malonyl-ACP with acetyl-CoA (rather than with acetyl-ACP). The protein sequence of the β-ketoacyl synthase-I encoded by the *fabB* gene has been deduced from the gene sequence; (v) β-ketoacyl-ACP reductase, which catalyses the NADPH-dependent reduction of the β-ketoacyl-ACP to (3R)-β-hydroxyacyl-ACP; (vi) β-hydroxyacyl-ACP dehydrase, which catalyses a *syn*-elimination with formation of *E*-2-enoyl-ACP; (vii) enoyl-ACP reductase. Two distinct enoyl reductases are known in *E. coli* having slightly different acyl chain length specificities, one NADH dependent and the other NADPH dependent; and finally (viii) palmityl-ACP thioesterase, which catalyses hydrolysis of the acyl thioester when the acyl chain reaches C_{16} in length.

The type-I FASs from yeast and mammals have a quite different architecture. In animal cells the component activities of FAS are found on a single polypeptide chain around 2500 amino acids in length. The native enzyme from chicken and rat is then a homodimer (M_r 500 000). Upon dissociating the polypeptide chains all the component activities are retained except that of the β-ketoacyl synthase. An elegant combination of protein biochemical and molecular genetic experiments have led to a detailed model for the architecture of the chicken FAS, in which the catalytic sites are arranged on a series of connected globular domains (Wakil 1989). Proteolytic cleavage of the chicken liver FAS leads initially to the release of three peptide fragments of M_r 127 000, 107 000 and 33 000 corresponding to domains I, II and III in the intact protein. The smallest, domain III, contains the COOH terminus of the protein and the thioesterase. Domain I contains the NH_2 terminus, the β-ketoacyl synthase, and a single active Ser-OH used by both acetyl and malonyl transacylases. Domain II (the reduction domain) contains the dehydrase and enoyl and β-ketoacyl reductases, as well as the acyl carrier site that connects the β-ketoacyl reductase to the thioesterase. This physical map based on proteolysis experiments is in accord with the predicted locations for the component activities in the polypeptide, based on an analysis of the entire sequence of chicken and rat cDNAs (see figure 4) (for a review see Hopwood & Sherman (1990)).

Clearly, for the decarboxylative condensation to occur the malonyl extender unit on the ACP must be positioned next to the acyl chain attached to the active site cysteine on the β-ketoacyl synthase. In the sequence of both chicken and rat FASs, the ACP and β-ketoacyl synthase are located almost at opposite ends of the polypeptide. Further important biochemical information has come from studies with alkylating agents. 1,3-Dibromopropane reacts rapidly and specifically with the FAS, inactivating only the condensing enzyme, by alkylating *both* the cysteine-SH on one polypeptide *and* the pantetheinyl-SH in the ACP *of the other chain*. These and related observations with DTNB-inhibited FAS led to the proposal that the two chains are arranged head to tail so that the reactive thiols of

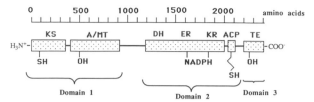

Figure 4. The location of activities along the type-I mammalian fatty acid synthase polypeptide deduced from the cDNA sequence (KS = β-ketoacyl synthase, A/MT = acetyl/malonyl transferase, DH = dehydrase, ER = enoyl reductase, ACP = acyl carrier protein domain, TE = thioesterase). The organization of activities into domains, as deduced by limited proteolysis, is also shown (Wakil, S. (1989) *Biochemistry* **28**, 4523).

the relevant ACP and condensing enzyme are juxtaposed (Wakil 1989). There are two full sites for fatty acid assembly. In this model each centre of palmitate synthesis derives the required component activities from both chains; one subunit contributes domain I whereas the second subunit contributes domains II and III.

The order of catalytic domains in the yeast FAS, deduced from cDNA sequence data, show some interesting differences to that found in the animal FAS (for a review see Hopwood & Sherman (1990)). The component activities are in this case distributed along two non-identical polypeptides. The α-subunit contains the acyl carrier site, the β-ketoacyl synthase and β-ketoacyl reductase, whereas the β-subunit contains domains for the remaining activities, including an FAD binding site in the enoyl reductase domain. In addition, there is no recognizable thioesterase domain, as expected because the end product is palmitoyl-CoA and not free palmitate. Thus the different end product of the FAS reflects a different combination of catalytic domains within the multifunctional protein, an observation that is of special interest in the context of the programming of polyketide synthase function.

4. THE STRUCTURE OF POLYKETIDE SYNTHASE COMPLEXES

Biochemical studies have proven a great success in the characterization of FASs from various organisms, however, the same is not true for PKSs from Streptomycetes. Several reasons for this can be found, including low levels of expression (typical for enzymes of secondary metabolism) under normal cellular conditions, and the very great problem of devising sensitive and specific assays applicable in cell free systems. Although the 6-methylsalicylic acid synthase has been purified from *Penicillium patulum*, and both chalcone synthase and resveratrole synthase have been purified from various plant species, no PKS active in polyketide antibiotic-producing Streptomycetes has yet been detected in an assay based on cell-free activity (for a review see Robinson (1989)). On the other hand, as a result of advances in molecular genetics, the genes encoding the biosynthetic pathways to several polyketide natural products have now been cloned from various *Streptomyces* species, including those for actino-

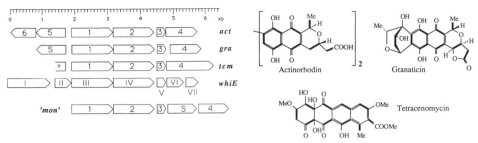

Figure 5. The organization of ORFs in the granaticin (*gra*), actinorhodin (*act*), tetracenomycin (*tcm*), *whiE* and '*mon*' polyketide synthase gene clusters. The deduced functions for these ORFs are: ORF1 + ORF2 (ORFIII + IV) = β-ketoacyl synthase; ORF3 (ORFV) = acyl carrier protein; ORF4 = cyclase/dehydrase (*gra*, *act*, and '*mon*'), cyclase/O-methyltransferase (*tcm*); ORF5 = ketoreductase; ORF6 = ketoreductase. See text for references.

rhodin, granaticin, oxytetracycline, tetracenomycin, daunorubicin, erythromycin, tylosin, spiramycin, carbomycin, avermectin, candicidin, as well as a spore pigment of unknown structure encoded by the *whiE* locus in *S. coelicolor* (for a review see Hopwood & Sherman (1990)). Two circumstances considerably facilitate the cloning of biosynthetic genes from Streptomycetes. One is that the genes of individual biosynthetic pathways are clustered on the host chromosome, and the second is that at least one resistance gene is always closely linked to the biosynthetic gene cluster. It is often relatively straightforward to clone an antibiotic resistance gene, and then to look in the flanking DNA for biosynthetic genes. An alternative, and more direct approach is by complementation of blocked mutants, and this was the method used to isolate biosynthetic genes for actinorhodin and tetracenomycin. From a genetic analysis of the DNA cluster, the region encoding the PKS genes can then usually be identified, and this upon sequencing leads directly to the primary structure(s) of the PKS protein(s).

The DNA for actinorhodin biosynthesis was the first antibiotic gene cluster to be cloned from a Streptomycete (Malpartida & Hopwood 1984). The DNA encoding the actinorhodin PKS was identified because it complemented mutants of the actinorhodin producing organism that were blocked in carbon chain assembly (*act*I and *act*III mutants). The sequence of this *act*I and *act*III DNA has been determined (Hopwood & Sherman 1990), but only that in the *act*III region has so far been reported in full (Hallam *et al.* 1988). Meanwhile, DNA sequences encoding PKSs required for granaticin and tetracenomycin biosynthesis have been described (Sherman *et al.* 1989; Bibb *et al.* 1989) and a computer assisted analysis revealed regions comprising a series of open reading frames (ORFs) (see figure 5) encoding discrete polypeptide chains. A third PKS cluster was cloned by complementing spore-pigment-deficient *S. coelicolor* mutants (*whiE*) (Davis & Chater 1990). When this DNA was sequenced a set of ORFs was again revealed, and several of the deduced protein sequences were closely related to those found in the actinorhodin and tetracenomycin PKS genes. The structure of the *whiE* spore pigment is presumably an aromatic polyketide, but its structure is presently unknown. Another putative polyketide synthase gene cluster has been

isolated from the monensin producer *S. cinnamonensis* by using the *act*I and *act*III DNA as probes to screen a genomic library (Arrowsmith 1990). Earlier it had been shown that DNA from *act*I and *act*III hybridizes to homologous sequences in genomic DNA from a large number of polyketide-producing Streptomycetes (Malpartida *et al.* 1987). Indeed, the homologous regions of DNA in granaticin- and oxytetracycline-producing Streptomycetes were later identified as the PKS genes involved in the production of these two antibiotics. The sequence of the *act*-homologous DNA from *S. cinnamonensis* has also been determined, and although its function is not yet known (it may be part of the monensin PKS cluster), a similar arrangement of encoded protein ORFs was identified (see figure 5).

A feature common to the sequenced *gra*, *act*, *tcm*, *whiE* and '*mon*' PKS gene clusters is a group of three ORFs, labelled ORF1-2-3, or ORFIII-IV-V, in figure 5. When corresponding ORFs from each cluster are compared, a very high end-to-end protein sequence similarity is observed. Moreover, the deduced protein sequences of ORFs 1 and 2 (ORFIII and IV) show a particularly high similarity with the *fabB* gene encoding the β-ketoacyl synthase-I from *E. coli*. In each case a potential active site Cys occurs only in ORF1 (ORFIII) (see figure 6). This fact, along with the possibility of translational coupling between these first two ORFs (because of overlapping 3' stop/5' start codons), suggests that the gene products are produced stoichiometrically to form a heterodimer, rather than being two distinct condensing enzymes. The deduced protein sequence of ORF3 (ORFV) in each cluster shows high end-to-end similarity to ACPs from bacteria and plants, as well as to ACP domains of animal type-I FAS. This includes a conserved Ser to which the cofactor is covalently attached (see figure 6). It is notable that plant chalcone and resveratrole synthases do not contain a covalently bound 4'-phosphopantetheine cofactor.

There is genetic evidence that ORF5 from the *act*III region encodes a ketoreductase, and the *gra* ORF5 and ORF6 as well as the '*mon*' ORF5 deduced protein sequences are very similar to this deduced *act*III gene product. These *act* ORF5, *gra* ORF5 (but not *gra* ORF6) and '*mon*' ORF5 sequences contain a typical consensus nucleotide binding motif Gly-Xaa-Gly-Xaa-Xaa-Ala (see figure 6). The *tcm* PKS cluster reveals no sequence homologous with these putative keto-

(a) β-ketoacyl synthase

```
      gra ORF1      G P V T M V S D G C T S G L D S V
      tcm ORF1      G P V T V V S T G C T S G L D A V
    whiE ORFIII     G P V Q T V S T G C T S G L D A V
    'mon' ORF1      G P V S T V S T G C T S G I D A V
      6-MSAS        G P S T A V D A A C A S S L V A I   (P. patulum)
      FAS fabB      G V N T S I S S A C A T S A H C I   (E. coli)
      FAS  rat      G P S I A L D T A C S S S L L A L
      FAS chicken   G P S L T I D T A C S S S L M A L
    ery ORFA (KS1)  G P A M T V D T A C S S S G L T A L
    ery ORFA (KS2)  G P A V T V D T A C S S S L V A L
```

(b) ACP

```
      gra ORF3      E E L G Y D S L A L M E S A
      tcm ORF3      Q D L G Y D S I A L L E I S
    whiE ORFV       D T F G L D S L G L L G I V
    'mon' ORF3      A L L G Y E S L A L L E T G
      6-MSAS        A D L G V D S V M T V T L R
      rat FAS       A D L G L D S L M G V E V R
    chicken FAS     A D L G L D S L M G V E V R
    S. erythraea    E D L G M D S L D L V E X V
    E. coli FAS     E D L G A D S L D T V E L V
    ery ORFA (ACP1) R D L G F D S M T A V D L R
    ery ORFA (ACP2) T E L G F D S L T A V G L R
```

(c) NAD(P) binding domains

```
      actIII              P V D V L V N N A G R P G G G A T A E L A D
      gra ORF5            T V D I L V N N A G R S G G G A T A E I A D
    'mon' ORF5            R I D V L V N N A G R S G G G V T A D L T D
      6-MSAS              L P R P E G T Y L I T G G L G V L G L E V A
    chick FAS¹⁸¹¹ KR      S C P P T K S Y I I T G G L G G F G L E L A
    rat FAS¹⁸⁰⁰ KR        F C P E H K S Y I I T G G L G G F G L E L A
    ery ORFA¹¹²¹ (KR1)    L E P L A G T V L V T G G I G A H L A R W L
    ery ORFA²⁵⁵⁸ (KR2)    S W E P A G T A L V T G G T G A L G G H V A
    DHFR  mouse          P L N C I V A V S Q N M G I G K N G D L P W
    chick FAS¹⁵⁹⁷ ER      K G E S V L I H S G S G G V G Q A A I A I A
    rat  FAS¹⁵⁸⁷ ER       H G E T V L I H S G S G G V G Q A A I S I A
```

Figure 6. Alignments of segments of deduced protein sequence: (*a*) around the presumptive active site cysteine in β-ketoacyl synthases; (*b*) around the active site serine of putative ACPs; (*c*) NAD(P)H binding domains in β-ketoacyl reductases and enoyl reductases.

reductases, consistent with the fact that no reductive step is required during polyketide assembly; all the β-placed oxygens are retained! There is also no ORF in the *whiE* cluster whose deduced protein sequence shows high similarity to those of *gra* and *act* ORF5.

The deduced protein sequences in ORF4 from *act*, *gra*, *tcm* and 'mon' show relatedness to each other, although the sequence similarities sometimes extend only over part of the deduced protein sequence. The *act*, *gra* and 'mon' ORF4 sequences are similar throughout the length of the protein, but each resembles *tcm* ORF4 only in the N-terminal half of the protein. Since *act* and *gra* ORF4 complement *act* VII mutants, deduced to be defective in cyclization and dehydration events, it has been suggested that *act* and *gra* ORF4s encode a bifunctional cyclase/dehydrase. In contrast, the *tcm* ORF4 protein seems to be a cyclase/O-methyltransferase because its C-terminal half resembles

bovine hydroxyindole-O-methyltransferase and *tcm* ORF4 complements *tcm* mutants lacking a specific O-methylation step (see Hopwood & Sherman 1990).

It is interesting that the deduced *gra* and *tcm* PKS clusters appear very similar at the protein level, despite the fact that polyketide chains are synthesized of different lengths and different patterns of folding. It seems likely that the folding of the polyketide chain and subsequent cyclization steps are catalysed by ORF4 and probably also by other proteins encoded outside the region sequenced so far. In any event, an important conclusion from this work is that these PKS genes for aromatic polyketides encode proteins most clearly resembling the type-II FAS from bacteria and plants, rather than being large multifunctional poly-peptides as seen in the type-I FAS and the 6-methylsalicylic acid synthase from *Penicillium patulum* (Beck *et al.* 1990).

Other studies on macrolide biosynthesis have recently furnished a remarkable view of macrolide PKS structure, which begin to provide an insight into how the synthesis of a macrolide is programmed at the molecular level. 6-Deoxyerythronolide B, the first isolable intermediate on the pathway to erythromycin A, is assembled from a propionyl-CoA starter unit and six methylmalonyl-CoA extender units (see figure 7). The genes encoding erythromycin biosynthesis are clustered in the genome of *Saccharopolyspora erythraea* and in the middle of the cluster lies the resistance gene *ermE*. Located about 12 kb downstream from this resistance genes is a DNA segment capable of complementing EryA mutants blocked in the biosynthesis of the erythronolide ring. This *eryAI* locus encodes the macrolide PKS. A further DNA segment homologous to *eryAI*, designated *eryAII*, was localized to a region about 35 kb downstream from *ermE*, and was also shown to encode genes for the macrolide PKS (Tuan *et al.* 1990).

The *eryAI* and *eryAII* DNA has now been sequenced and this has provided crucial information, not only about the size, but also about the probable functions of ORFs encoded in the *ery* PKS cluster. The sequence of one such ORF (ORFA) extending over 9.5 kb of DNA has been reported by Leadlay's group (Cortes *et al.* 1990). The deduced gene product is predicted to

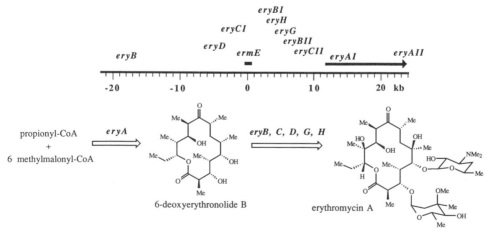

Figure 7. The erythromycin biosynthetic gene cluster, showing relative locations of the resistance gene (*ermE*) and biosynthetic genes (*eryA-eryH*) (Weber *et al.* 1990).

Phil. Trans. R. Soc. Lond. B (1991)

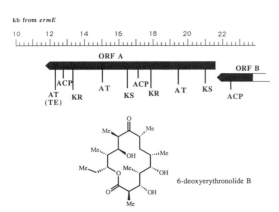

Figure 8. An open reading frame in the *ery* cluster (*eryAl*) encoding a multifunctional polypeptide, where the deduced activities are: AT(TE) = acyl transferase (thioesterase), ACP = acyl carrier protein domain, KR = ketoreductase, KS = β-ketoacyl synthase (Cortes *et al.* 1990).

contain 3178 amino acids. Upon comparison with available protein-sequence databases, nine separate portions of the deduced protein are found to be very similar to active-site sequences found in the constituent catalytic activities of known FASs and PKSs. The deduced activities for various regions of this predicted protein are shown in figure 8, and include in sequence: AT(or 'TE')-ACP-KR-AT-KS-ACP-KR-AT-KS (see figure 8 for definitions of symbols). A comparison of putative active site residues in this ORF with those found in corresponding components of other FAS and PKS complexes is shown in figure 6. These results reinforce the earlier conclusion that this portion of the *ery* cluster encodes a PKS, and show also that the bacterial PKSs are not necessarily complexes of monofunctional proteins, a trend evident with the aromatic PKSs.

In a parallel investigation Katz's group have also sequenced further downstream and found two additional large ORFs of comparable size to that described by Leadlay. Each of these ORFs again appears to encode a single large multifunctional protein containing multiple copies of units comprising AT, ACP, KS, KR activities, and one ORF contains also one set of dehydrase and enoyl reductase activities (L. Katz; personal communication). It seems, therefore, that the erythronolide PKS may comprise three large multienzyme complexes, possessing all the activities required to catalyse six rounds of chain extension and modification. With this information in hand it becomes possible to speculate as to how the individual steps in 6-deoxyerythronolide biosynthesis might be catalysed by the deduced activities encoded in the three *ery* ORFs. Katz has proposed a 'module-hypothesis' in which each polypeptide carries the functions for two rounds of chain elongation and modification. The DNA sequence for each round is called a 'module'. The deduced activities encoded in ORF A shown in figure 8 would comprise the last two modules required to add the final two propionate units and complete the synthesis of 6-deoxyerythronolide A. There are apparently two condensing enzymes and two keto-reductases, and at the C-terminus of the deduced protein is a sequence showing similarity to known

thioesterases, which might therefore catalyse the final act of macrolide ring formation. This hypothesis implies that each of the deduced activities KS, KR, DH and ER is used once during the synthesis of each molecule of 6-deoxyerythronolide, and consistent with this is the total number of such activities; $6 \times KS$, $5 \times KR$, $1 \times DH$, $1 \times ER$. There are also multiple copies of deduced ACP domains and AT activities. This leaves sufficient scope for differences in stereospecificity to arise amongst the ketoreductases, for example, or among the acyl transferases ((*R*)- versus (*S*)-methylmalonyl-CoA (Hutchinson 1983)). There are no real clues yet as to how the growing acyl chain is transferred between active sites, and between the three different polypeptide chains. Leadlay has suggested that the two halves of the ORFA gene product might fold back on each other so as to bring the predicted β-ketoacyl synthase active sites into close proximity to the ACP domains (Cortes *et al.* 1990).

The remarkable picture emerging from these studies is one of a colinearity between the biochemical steps in macrolide assembly and the genetic order of ORFs encoding the requisite multifunctional proteins in the genome of the producing organism. Whether this picture has been interpreted correctly will no doubt be tested by future biochemical experiments. Although one swallow does not make a summer, should this correlation prove to be a general one among macrolide and polyether PKS complexes, then an important step forward will have been made in understanding the programming of these polyketide pathways at the biochemical and genetic levels. The prospect also arises of using rDNA methods to engineer new biosynthetic pathways. Protein engineering experiments might, for example, involve swapping modules or ORFs between polyketide pathways in order to make homologues of known antibiotics, or bringing modules or ORFs into novel combinations so that the host microorganism is programmed to biosynthesize entirely new classes of natural products. Alternatively, modules or simply portions of modules might be deleted or disrupted to generate new end products, or release intermediates in the assembly process. Although the prospects seem good, determining where the limits lie for the rational manipulation of polyketide biosynthesis will probably require a far more extensive knowledge of these processes, at all levels.

The author thanks Professor David Hopwood for valuable discussions, and Dr Leonard Katz and co-workers for providing information on the *ery* PKS prior to publication.

REFERENCES

Ajaz, A. A., Robinson, J. A. & Turner, D. L. 1987 Biosynthesis of the polyether ionophore antibiotic monensin A: assignment of the carbon-13 and proton NMR spectra of monensin A by 2D spectroscopy. Incorporation of oxygen-18 labelled molecular oxygen. *J. chem. Soc. Perkin Trans. I* 27–36.

Arrowsmith, T. J. 1990 Characterisation of putative polyketide synthase genes from *Streptomyces cinnamonensis*. Ph.D. thesis, University of Southampton.

Beck, J., Ripka, S., Siegner, A., Schiltz, E. & Schweizer, E. 1990 The multifunctional 6-methylsalicylic acid synthase gene of *Penicillium patulum*. Its gene structure relative to that of other polyketide synthases. *Eur. J. Biochem.* **192**, 487–498.

Bibb, M. J., Birö, S., Motamedi, H., Collins, J. F. & Hutchinson, C. R. 1989 Analysis of the nucleotide sequence of the *Streptomyces glaucescens tcmI* genes provides key information about the enzymology of polyketide antibiotic biosynthesis. *EMBO J* **8**, 2727–2736.

Birch, A. J. 1957 Biosynthetic relations of some natural phenolic and enolic compounds. *Prog. Chem. Org. Nat. Prod.* **14**, 186–216.

Birch, A. J. 1967 Biosynthesis of polyketides and related compounds. *Science, Wash.* **156**, 202–206.

Birch, A. J. & Donovan, F. W. 1953 Studies in relation to biosynthesis I. Some possible routes to derivatives of orcinol and phloroglucinol. *Aust. J. Chem.* **6**, 360–368.

Brockmann, H. & Henkel, W. 1951 Pikromycin ein bitter schmeckender Antibioticum aus Actinomyceten *Chem. Ber.* **84**, 284–288.

Cane, D. E., Liang, T.-C. & Hasler, H. 1982 Polyether Biosynthesis 2. Origin of the oxygen atoms of monensin A. *J. Am. chem. Soc.* **104**, 7274–7281.

Cane, D. E., Celmer, W. D. & Westley, J. W. 1983 Unified stereochemical model of polyether antibiotic structure and biogenesis. *J. Am. chem. Soc.* **105**, 3594–3600.

Cane, D. E. & Ott, W. R. 1988 Macrolide Biosynthesis 5. Intact incorporation of a chain-elongation intermediate into nargenicin. *J. Am. chem. Soc.* **110**, 4840–4841.

Cane, D. E. & Yang, C.-C. 1987 Macrolide biosynthesis 4. Intact incorporation of a chain-elongation intermediate into erythromycin. *J. Am. chem. Soc.* **109**, 1255–1257.

Cortes, J., Haydock, S. F., Roberts, G. A., Bevitt, D. J. & Leadlay, P. F. 1990 An unusually large multifunctional polypeptide in the erythromycin-producing polyketide synthase of *Saccharopolyspora erythraea*. *Nature, Lond.* **348**, 176–178.

Davis, N. K. & Chater, K. F. 1990 Spore colour in *Streptomyces coelicolor* A(3)2 involves the developmentally regulated synthesis of a compound biosynthetically related to polyketide antibiotics. *Molec. Microb.* **4**, 1679–1691.

Fulco, A. J. 1983 Fatty acid metabolism in bacteria. *Prog. Lipid Res.* **22**, 133–160.

Hallam, S. E., Malpartida, F. & Hopwood, D. A. 1988 Nucleotide sequence, transcription and deduced function of a gene involved in polyketide antibiotic synthesis in *Streptomyces coelicolor*. *Gene* **74**, 305–320.

Hopwood, D. A. & Sherman, D. H. 1990 Molecular genetics of polyketides and its comparison to fatty acid biosynthesis. *A. Rev. Genet.* **24**, 37–66.

Hutchinson, C. R. 1983 Biosynthetic studies of macrolide and polyether antibiotics. *Acct. chem. Res.* **16**, 7–14.

Malpartida, F. & Hopwood, D. A. 1984 Molecular cloning of the whole biosynthetic pathway of a *Streptomyces* antibiotic and its expression in a heterologous host. *Nature, Lond.* **309**, 462–464.

Malpartida, F., Hallam, S. E., Kieser, H. M., Motamedi, H., Hutchinson, C. R., Butler, M. J., Sugden, D. A., Warren, M., McKillop, C., Bailey, C. R., Humphreys, G. O. & Hopwood, D. A. 1987 Homology between *Streptomyces* genes coding for synthesis of different polyketides used to clone antibiotic biosynthesis genes. *Nature, Lond.* **325**, 818–820.

Robinson, J. A. 1988 Enzymes of secondary metabolism in microorganisms. *Chem. Soc. Rev.* **17**, 383–452.

Sherman, D. H., Malpartida, F., Bibb, M. J., Kieser, H. M., Bibb, M. J. & Hopwood, D. A. 1989 Structure and deduced function of the granaticin-producing polyketide synthase gene cluster of *Sterptomyces violaceoruber* Tü22. *EMBO J.* **8**, 2717–2725.

Takano, S., Sekiguchi, Y., Shimazaki, Y. & Ogasawara, K. 1989 Stereochemistry of the proposed intermediates in the biosynthesis of mycinamicins. *Tetrahedron Lett.* **30**, 4001–4002.

Tuan, J. S., Weber, J. M., Staver, M. J., Leung, J. O., Donadio, S. & Katz, L. 1990 Cloning of genes involved in erythromycin biosynthesis from *Saccharopolyspora erythraea* using a novel actinomycete-*E. coli* cosmid. *Gene* **90**, 21–29.

Volpe, J. J. & Vagelos, P. R. 1973 Saturated fatty acid biosynthesis and its regulation. *A. Rev. Biochem.* **42**, 21–60.

Wakil, S. 1989 Fatty acid synthase, a proficient multifunctional enzyme. *Biochemistry* **28**, 4523–4530.

Weber, J. M., Leung, J. O., Maine, G. T., Potenz, R. H. B., Paulus, T. J. & DeWitt, J. P. 1900 Organisation of a cluster of erythromycin genes in *Saccharopolyspora erythraea*. *J. Bact.* **172**, 2372–2383.

Woodward, R. B. 1957 Struktur und Biogenese der Makrolide; Eine neue Klasse von Naturstoffen. *Angew. Chem.* **69**, 50–58.

Yue, S., Duncan, J. S., Yamamoto, Y. & Hutchinson, C. R. 1987 Macrolide biosynthesis. Tylactone formation involves the processive addition of three carbon fragments. *J. Am. chem. Soc.* **109**, 1253–1255.

To build an enzyme...

JEREMY R. KNOWLES

Departments of Chemistry and Biochemistry, Harvard University, Cambridge, Massachusetts 02138, U.S.A.

SUMMARY

The structural components that lead to enzyme function are discussed for one simple enzyme-catalysed reaction: that mediated by triosephosphate isomerase. First, the recognition and binding of the substrates' phospho group is seen to involve four main-chain –NH– hydrogen bonds, two of which are positioned at the positive end of a short α-helix aimed precisely at the phospho group and interact with the three peripheral phospho group oxygens. Second, the chemical steps (of substrate enolization) are shown to require both base and general acid catalysis. The identity and the positioning of the base, a carboxylate group, nicely fulfils the expectations both of mechanistic economy and of stereoelectronics. The identity of the general acid is shown by Fourier transform infrared and by ^{15}N nuclear magnetic resonance (NMR) to be a neutral imidazole group, lying between the two substrate oxygens. The positioning of the ring is ideal, but its protonation state is unexpected. Thus the pK_a of this histidine side-chain is < 4.5, lowered from 6.5 (the value in the denatured protein) by its position at the positive end of another well-aimed α-helix. Third, the need for enzymes to provide kinetic barriers to the loss of reaction intermediates from the active site is emphasized. Triosephosphate isomerase achieves this sequestration of the reaction intermediate by using a flexible loop of the protein, and thus improves the efficiency of the catalysed transformation.

1. INTRODUCTION

There are two views of the catalytic power of enzymes. The first is that the catalytic rates achieved by enzymes are extraordinary, that enzymes represent the rare end products of an extensive search through protein sequence 'space' (Maynard-Smith 1970), and, perhaps, that they use mechanistic and kinetic devices that have not yet been recognized by physical-organic chemists. The second view is less awesome, and is that each of today's enzymes represents one of a large number of more-or-less equally effective possible solutions to the catalytic problem, each of these solutions using functional groups and other catalytic elements in ways that are consonant with the expectations of mechanistic organic chemistry. In this paper, we examine the validity of the second view by dissecting the catalytic apparatus of an enzyme that mediates a chemically simple reaction in the glycolytic pathway: triosephosphate isomerase.

Triosephosphate isomerase catalyses the interconversion of the two triose phosphates, dihydroxyacetone phosphate and *R*-glyceraldehyde 3-phosphate. The overall reaction simply involves the enolization of substrate to the enediol (or enediolate) intermediate, which then re-ketonizes to form the product (Rieder & Rose 1959; Rose 1962). The enzyme thus mediates two enolizations, back to back, as shown in figure 1. Already in figure 1 we have shown two catalytic groups, a base B to abstract the carbon-bound proton and a general acid H-A to assist in this process. Before any consideration of active site chemistry, however, let us look at the first step of the catalysed process, in which the substrate is recognized and binds to the active site.

2. SUBSTRATE BINDING

In the case of the two triose phosphates, the obvious 'handle' for substrate binding is the phosphate ester, which indeed seems to dominate the recognition process. Thus while many phosphate ester analogues of the substrates are competitive inhibitors of the enzyme (these range from molecules that bind relatively weakly such as glycerol 1-phosphate and 3-phosphoglycerate (Lambeir *et al.* 1987) to more impressive inhibitors such as phosphoglycolate and phosphoglycolohydroxamate (Collins 1974)). Any change in the phospho group itself (as in dihydroxyacetone sulphate (Belasco *et al.* 1978)) results in molecules that are not recognized by the enzyme and do not bind to it. From nuclear magnetic resonance (NMR) experiments and pH-variation studies (Campbell *et al.* 1978; Belasco *et al.* 1978), it is known that the phospho group binds as a dianion, and it is instructive to see how the enzyme creates a binding locus for this group. This is shown in figure 2. No cationic amino acid side chains (of lysine, arginine or histidine) are involved, the peripheral oxygens of the phospho group being held by at least four main-chain –NH– hydrogen bonds (Davenport 1986; Lolis & Petsko 1990). As Pauling pointed out many years ago (Pauling 1960), a main-chain –NH– bond is 'worth' about half a charge, and we may expect that the arrangement shown in figure 2

Phil. Trans. R. Soc. Lond. B (1991) **332**, 115–121
Printed in Great Britain

[9]

115

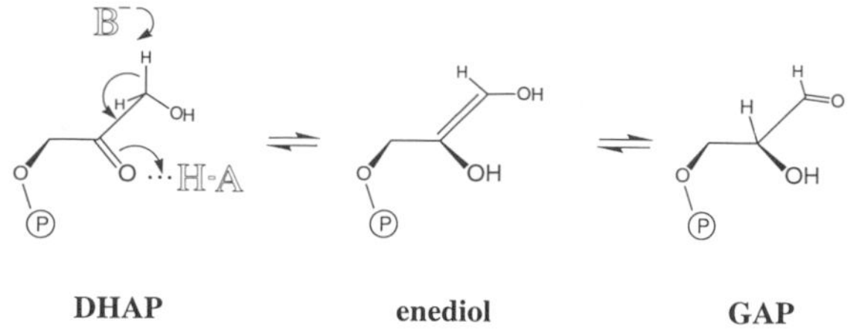

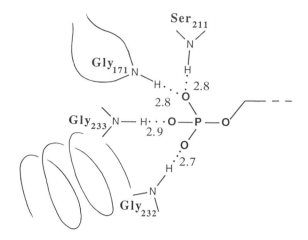

DHAP   **enediol**   **GAP**

Figure 1. The mechanistic pathway of the reaction catalysed by triosephosphate isomerase. B is a catalytic base, and H–A is a catalytic electrophile. DHAP is dihydroxyacetone phosphate, and GAP is *R*-glyceraldehyde 3-phosphate.

effectively balances the dianionic charge of the phospho group. Yet if electrostatic neutrality was all that was important, why did the enzyme not use lysine or arginine side chains? A persuasive answer to this question lies in the evident need (which will become apparent below) for precision in the positioning of substrate with respect to the catalytic groups. The side chains of lysine and arginine are both flexible, and essentially all side chains in proteins have greater freedom and mobility than the atoms of the main chain. Indeed it seems unlikely that a binding locus of such constraining exactness as that shown in figure 2 could easily be created by using the floppier side chains of lysine or arginine. (These views are strongly tinged with hindsight, of course, for there exist many examples where phospho group binding sites do contain lysine, arginine, or histidine side chains (see, for example, Cotton *et al.* (1979); Sowadski *et al.* (1985); Luecke & Quiocho (1990)). Yet it is true that even in these cases, the opportunity for multiple hydrogen bonds almost always ties the side chain cation back to the main-chain framework.)

There are two other features of figure 2 that deserve comment. Two of the main-chain hydrogen bonds derive from glycine residues 232 and 233 that lie at the positive end of a short α-helix that is beautifully aimed at the substrates' phospho group (figure 3*b*). It is hard to imagine that the precision of this alignment is accidental, and even if the effect of an α-helix dipole (Hol 1985) is, as has been suggested, predominantly because of the alignment of the hydrogen bond donors and acceptors in the final turn, the arrangement appears to be a particularly sturdy one. The second important feature of figure 2 is that glycine-171 lies on a flexible loop of residues (from 166 to 176) that closes down over the active site when the substrate binds. The mechanistic function of this loop movement and the catalytic consequence of the recruitment of an additional group to bind the substrate, are discussed later. Suffice it here to conclude that the substrates' phospho group seems to be nicely accommodated by the constellation of hydrogen bond donors shown in figure 2.

3. THE FIRST CHEMICAL STEP

The covalency changes that produce the enediol(ate) intermediate occur within the first-formed Michaelis complex. Enolizations are both acid- and base-catalysed, and the kinetic benefit from the concerted

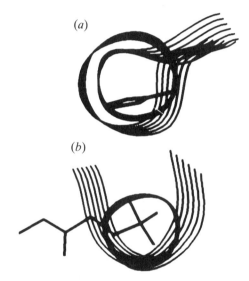

Figure 2. A schematic representation of the phospho group binding locus of triosephosphate isomerase. Gly-232 and Gly-233 lie at the positive end of a short α-helix that is aimed precisely at phosphorus, and Gly-171 lies on the flexible loop that closes down over the bound substrate. The numbers beside the indicated hydrogen bonds are N-to-O distances, in Å.

Figure 3. Illustrating the aim of the two α-helices that point into the active site. (*a*), looking down the helix that is aimed at the imidazole ring of histidine-95; (*b*), looking down the helix that is aimed at the phospho group of bound substrate. The coordinates are from the crystal structure of the yeast enzyme containing bound phosphoglycolohydroxamate (Davenport 1986).

action of both has been well documented (Hegarty & Jencks 1975). Certainly the organic chemist would be drawn to propose a base and a general acid arranged as shown in figure 1. It is, therefore, gratifying that from the high resolution crystal structures of triosephosphate isomerase containing bound substrate or substrate-like inhibitors (Banner *et al.* 1975; Alber *et al.* 1981; Lolis *et al.* 1990; Lolis & Petsko 1990), a base (the carboxylate of glutamate-165) and a possible acid (the imidazole of histidine-95) are nicely placed to mediate the enolization. The nearer carboxylate oxygen of glutamate-165 is 2.8 Å† and 3.4 Å from C-1 and C-2 of the substrate, respectively, thus poised for

† $1 \text{ Å} = 10^{-10} \text{ m} = 10^{-1} \text{ nm}$.

the abstraction either of the C-1 pro-*R* proton of dihydroxyacetone phosphate or of the C-2 proton of glyceraldehyde phosphate (whichever triose phosphate substrate binds). The use by the enzyme of a bidentate base seems particularly appropriate, considering the reaction to be catalysed (figure 1). The base is properly positioned in stereoelectronic terms too, being orthogonal to (what becomes) the plane of the enediol (that is, the plane defined by O-1, C-1, C-2 and O-2) as required by Corey & Sneen (1956), and using the *syn* orbitals of the carboxylate, following Gandour (1981). From all chemical points of view, therefore, the identity and position of the catalytic base in triosephosphate isomerase seems ideal. Unsurprisingly, such an attractive arrangement is not unique. One particularly well-studied parallel is the enzyme Δ^5-3-ketosteroid isomerase (Kuliopulos *et al.* 1989), where the carboxylate of an aspartate residue is used analogously to effect the abstraction and replacement of carbon-bound protons (in a 1,3 shift, rather than the 1,2 relation in triosephosphate isomerase).

From the crystal structure of the liganded isomerase, it is clear that the imidazole ring of histidine-95 is an excellent candidate for the general acid H-A. The N^ϵ of this residue is almost equidistant from the substrate oxygens O-1 and O-2 (the distances are 2.8 Å and 2.9 Å, respectively), well placed in the plane of what will be the enediol and at a distance to make appropriately strong hydrogen bonds to substrate. Once again the positioning seems to be ideal, according nicely with the prejudices of the physical-organic chemist who would like to catalyse an enolization. Just as early work by chemical modification had anticipated the existence of glutamate-165 as the isomerase's catalytic base (Hartman 1968; Waley *et al.* 1970; de la Mare *et al.* 1972), early chemical and spectroscopic experiments (Webb & Knowles 1975; Belasco & Knowles 1980) had suggested the existence of a general acid that polarizes the carbonyl group of enzyme-bound substrate. Thus the infrared stretching frequency of the carbonyl group of dihydroxyacetone phosphate is moved by 19 cm^{-1} to lower frequency when this substrate binds to the isomerase (Belasco & Knowles 1980). That this polarizing shift is due to the imidazole ring of histidine-95 has recently been supported by infrared studies on the substrates bound to two mutant isomerases H95N and H95Q, in which histidine-95 has been changed either to asparagine or to glutamine. Neither of these mutant enzymes (the catalytic potency of which are 10^4- and 10^2-fold less than the wild-type enzyme, respectively, Nickbarg *et al.* (1988); Blacklow & Knowles (1990)) polarizes the substrate's carbonyl group (Komives *et al.* 1991), and these data are entirely consistent with the indications from the protein crystal structure that histidine-95 is the catalytic electrophile.

At this point, the chemical reader would naturally presume that histidine-95 would be in its protonated imidazolium form at the pH-values where the enzyme is catalytically active, so as to act as an effective electrophile. The pK_a of a histidine side chain is normally somewhat less than 7, and the imidazolium cation is (with p$K_a \approx 7$) obviously a much stronger

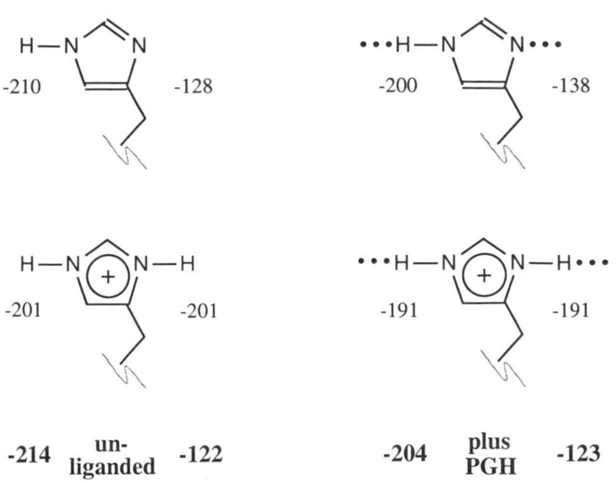

Figure 4. Typical ^{15}N NMR chemical shifts for imidazole nitrogens in different environments. All values are upfield (negative) from 1M HNO$_3$. The values in bold type are those observed for the double mutant (H103Q·H185Q) yeast triosephosphate isomerase containing ring-labelled [^{15}N$_2$]-histidine-95, either alone ('unliganded') or in the presence of saturating levels of phosphoglycolohydroxamate ('plus PGH').

electrophile than neutral imidazole (with p$K_a \approx 14$). Yet these comforting chemical statements were not consistent with the crystallographic results, which suggested that the imidazole ring of histidine-95 is neutral (Browne *et al.* 1976). This conclusion was based upon the fact that the distal imidazole nitrogen, N^δ, lies within hydrogen bond distance of the main-chain –NH– group of residue 97. Clearly if N^δ acts as a hydrogen bond acceptor, the imidazole ring cannot be protonated.

To resolve the question, we have investigated the protonation (and hydrogen bonding) state of the two nitrogens of the imidazole ring of histidine-95, by ^{15}N NMR. To simplify the spectrum and to eliminate the need for resonance assignment, we took the yeast enzyme (which has three histidine residues, at positions 95, 103, and 185) and replaced the two histidines (at positions 103 and 185) in which we had no interest. The resulting double-mutant enzyme H103Q·H185Q is mechanistically and kinetically indistinguishable from the wild type, and the only histidine in the molecule is that at the active site. The mutant enzyme was produced in an auxotrophic *his*$^-$ *Escherichia coli* host, grown on singly or doubly [^{15}N]-labelled histidine. As can be seen from figure 4, the chemical shift of the ^{15}N resonances of imidazole nitrogens is exquisitely sensitive to protonation state and participation in hydrogen bonds (Bachovchin 1986) and has allowed the nature and environment of the side chain of histidine-95 to be probed (P. Lodi, unpublished data). A typical spectrum from the doubly [^{15}N]-labelled yeast isomerase is shown in figure 5. From these data, it is evident that, contrary to mechanistic expectation, the imidazole ring of histidine-95 is not protonated at neutral pH. Indeed, as is shown by the insert in figure 5, the ring remains unprotonated down at least to pH 5. The pK_a of histidine-95 must be less than 4.5! Also evident from

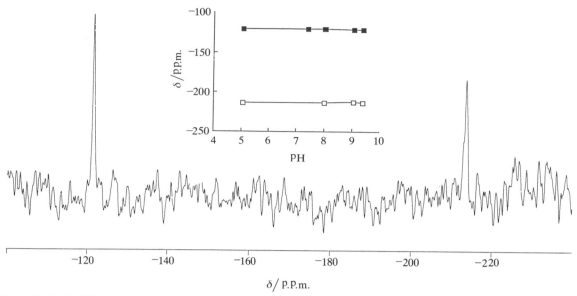

Figure 5. Partial ^{15}N NMR spectrum of unliganded H103G·H185Q yeast triosephosphate isomerase containing doubly ring-labelled [^{15}N$_2$]histidine-95, at pH 9.4. Inset: pH-variation of the chemical shifts for the two resonances observed in the ^{15}N NMR spectrum.

the experimental results shown (in bold face) in figure 4, is the fact that when the substrate analogue phosphoglycolohydroxamate binds, the N$^{\varepsilon}$ of histidine-95 becomes a hydrogen-bond donor. We may conclude, based upon this NMR result (P. Lodi, unpublished data) and the crystal structure of the yeast enzyme with bound inhibitor (Davenport 1986; Lolis & Petsko 1990), that a strong hydrogen bond (stronger, that is, than to any water molecules that occupy the active site in the absence of ligand) is indeed formed between N$^{\varepsilon}$ and the substrate analogue's carbonyl oxygen.

Two questions are raised by these results. First, how is the pK_a of histidine-95, which we have determined to be 6.5 in the denatured protein, lowered by more than two units in the native enzyme? Inspection of the structure provides an immediate answer, that is shown in figure 3a. The imidazole ring of histidine-95 lies at the positive end of a short α-helix, and N$^{\delta}$ is within hydrogen-bonding distance of two main-chain –NH– groups, of residues 96 and 97 (at 2.7 Å and 2.9 Å, respectively). As has been noted before (Hol 1985; Perutz *et al.* 1985), and most clearly shown by the work of Fersht and his group on barnase (Sali *et al.* 1988), such a local environment readily accounts for the lowering of the pK_a of a histidine residue. The second question is more difficult and more teleological: why should the enzyme go to such lengths to use imidazole, a lesser electrophile, in preference to imidazolium? (This question presumes, pending the completion of analogous but technically more difficult experiments with substrate itself, that DHAP and phosphoglycolo-hydroxamate will have similar effects on the ^{15}N NMR of the labelled isomerase.) This issue cannot be usefully analysed without more information, though several possibilities can be mentioned. Thus (i) the juxtaposition of two oppositely charged catalytic groups (a carboxylate and an imidazolium) might lead to an ion pairing interaction that would preclude the proper positioning of these groups for catalysis; (ii) the

existence of a local imidazolium might lower the basicity of the carboxylate and thus slow the enolization reaction; (iii) since the pK_a of imidazole is known to follow (some seven units away) the pK_a of imidazolium (Bruice & Schmir 1958; Walba & Isensee 1961), it may be that the enzyme lowers the pK_a of imidazole to a value close to that of the enediol intermediate, thus allowing for very rapid proton transfers between O-1 and O-2, and (iv) there may be some unsuspected mechanism by which an imidazole can catalyse the transfer of protons between the oxygens of a *cis*-enediolate. Whatever the reasons behind the enzyme's use of a neutral imidazole as the electrophile, it is clear that this represents a gap in our understanding.

4. THE NEED FOR SEQUESTRATION OF INTERMEDIATES

Any enzyme-catalysed reaction that involves a reaction intermediate *en route* from substrate to product, faces the problem of avoiding the loss of the intermediate from the active site. As shown in figure 6a, the substrate (or the product) can, when enzyme-bound, partition in two ways: forward, involving the chemical changes of the reaction, or backward, involving the diffusive loss from the active site. These alternatives are necessary and sufficient for all one-substrate enzymic processes. (Extension of these arguments to multi-substrate reactions having either random or ordered binding, is straightforward.) Consider, however, the reaction intermediate, which, as shown in figure 6b, must partition in two directions only, each of which involves the chemical changes of the catalysed reaction: forward to GAP or backward to DHAP. The notational third route, of departure from the active site, must be blocked, so as to ensure a stoichiometric 'throughput' of substrate to product. In the case of triosephosphate isomerase, for example, it is known that in free solution the enediol intermediate

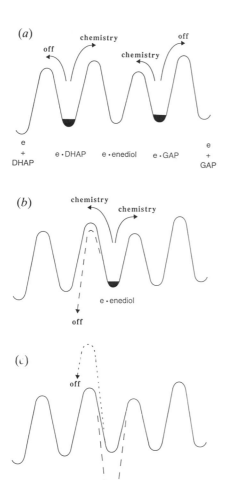

Figure 6. Free energy profiles for the action catalysed by triosephosphate isomerase, showing the need for sequestration of the reaction intermediate. (*a*), how the complexes of either of the substrates dihydroxyacetone phosphate (DHAP) or *R*-glyceraldehyde 3-phosphate (GAP) with the enzyme (e) may partition in two ways, either forward (to involve chemistry) or back (to fall off the enzyme); (*b*), the complex of the reaction intermediate (enediol) with the enzyme (e), must partition only forward or back involving chemistry, and not fall off the enzyme; (*c*), showing the two ways in which the enzyme can avoid losing its grip on the reaction intermediate, either by the relative thermodynamic stabilization of the enzyme–enediol complex, or by the creation of a kinetic barrier to the loss of the enediol from the active site.

decomposes (to methylglyoxal and P_i) about 100 times faster than it reprotonates on carbon to give either dihydroxyacetone phosphate or glyceraldehyde phosphate (Richard 1984). Triosephosphate isomerase must not, therefore, lose its grip on the enediol(ate) intermediate, and this problem of intermediate sequestration is a general one for essentially all enzymes. (Only for those transformations involving no reaction intermediates, such as the S_N2-like reactions of *S*-adenosylmethionine or most phosphokinase reactions, is this question unimportant.)

The difficulty of ensuring that a reaction intermediate is retained by the enzyme is not, of course, overcome by simply lowering the free energy of the enzyme–intermediate complex relative to the complexes with substrate or with product. While such an arrangement would certainly prevent intermediate

loss, it would produce an inefficient catalyst that, in free energy terms, would fall into the well of the enzyme–enediol complex and stay there (figure 6*c*, dashed line). What an enzyme must do is to create a kinetic barrier for the loss of the intermediate (figure 6*c*, dotted line) that is much higher than the kinetic barriers for the loss of substrate or product. For an unstable reaction intermediate that is of much higher free energy than the substrate or product, this can be effected simply by preferential binding of the intermediate. In this way, the enzyme can ensure that the intermediate partitions only between two chemical paths, and that there is no opportunity for loss from the active site. How this is achieved by triosephosphate isomerase is described below.

In common with many enzymes, triosephosphate isomerase has a flexible loop of ten amino acid residues (that have recently been more aptly described as a lid: Joseph *et al.* (1990)) that appears to close over the substrate during the catalytic act. Crystallographic analysis had first indicated that this lid adopts a 'closed' position when substrate or a substrate analogue occupies the active site (Banner *et al.* 1975; Alber *et al.* 1981), and it was then conjectured that the conformation change (which amounts to induced fit (Koshland 1959), even if new catalytic functionalities are not thus recruited) could prevent loss of the enediol(ate) intermediate from the enzyme. It had been early recognized that the enzyme must (and does) bind its substrates so as to keep the phospho group more or less in the plane of the enediol(ate). Such a conformation stereoelectronically disfavours the decomposition of the intermediate to methylglyoxal and P_i, a reaction that, as noted above, is normally much faster than C-protonation (Richard 1984). This facile decomposition makes it especially important that the intermediate not be released. (The consequences of intermediate loss are not, of course, always disastrous, for the intermediate may be transiently stable (e.g. the aminoacyladenylate from tRNA synthetases, or hydroxyethylthiamin pyrophosphate), or the intermediate may be very unstable, but collapse nonenzymically to the substrate and product anyway (e.g. the carbanionic intermediate from amino acid racemases, or the dienolate made by Δ^5-3-ketosteroid isomerase).)

To test these ideas, we deleted four amino acid residues from the lid of triosephosphate isomerase, choosing a segment that minimized any non-local changes in the enzyme's structure (Pompliano *et al.* 1990). The abbreviated loop (which lacks, *inter alia*, glycine-171, see figure 2) can no longer encapsulate the substrate, and model building suggests that the active site of this mutant enzyme is at all times open to solvent. The purified 'loopless' mutant enzyme is a much less effective enzyme. Values of k_{cat} are some 10^5-fold lower than the wild-type, although K_m values rise by less than tenfold. Evidently the transition states for the chemical steps have been most sharply affected, and from the much weaker binding of an analogue of the enediol(ate) intermediate, it seems that the enzyme–intermediate complex is relatively much less stable, too (Pompliano *et al.* 1990). The first and most obvious role for the lid residues, then, is preferentially to stabilize the enediol(ate) intermediate (as distinct from the substrate or the product) and the two enolization transition states that flank it. Yet the second function (which is relevant to the question of intermediate sequestration discussed above), to provide a kinetic barrier to the loss of the reactive intermediate, is also evident. Indeed, when the stoichiometry of substrate–product conversion was investigated, we found that for six molecules of glyceraldehyde phosphate consumed by the loopless mutant isomerase, only one molecule of dihydroxyacetone phosphate was formed. The other five molecules ended as methylglyoxal and P_i, the presumed consequence of the loss of the intermediate from the enzyme and its rapid decomposition in free solution.

It thus appears that triosephosphate isomerase has neatly and economically used a flexible loop of the protein, both to speed the reaction by preferential stabilization of the reaction intermediate and of the transition states that lead to it, and to maximize the conversion of substrate to product by creating a kinetic barrier to the loss of the intermediate from the active site.

The contributions of many collaborators, especially Steve Blacklow, Elizabeth Komives, Patricia Lodi, Elias Lolis, Elliott Nickbarg, Greg Petsko, Anusch Peyman, David Pompliano and Ron Raines, and the financial support of the National Institutes of Health, is gratefully acknowledged.

REFERENCES

Alber, T., Banner, D. W., Bloomer, A. C., Petsko, G. A., Phillips, D. C., Rivers, P. S. & Wilson, I. A. 1981 On the three-dimensional structure and catalytic mechanism of triosephosphate isomerase. *Phil. Trans. R. Soc. Lond.* B **293**, 169–171.

Bachovchin, W. W. 1986 ^{15}N NMR spectroscopy of hydrogen-bonding interactions in the active site of serine proteases: evidence for a moving histidine mechanism. *Biochemistry* **25**, 7751–7759.

Banner, D. W., Bloomer, A. C., Petsko, G. A., Phillips, D. C., Pogson, C. I., Wilson, I. A., Cowan, P. H., Furth, A. J., Milman, J. D., Offord, R. E., Priddle, J. D. & Waley, S. G. 1975 Structure of chicken muscle triosephosphate isomerase determined crystallographically at 2.5 Å resolution using amino acid sequence data. *Nature, Lond.* **255**, 609–614.

Belasco, J. G., Herlihy, J. M. & Knowles, J. R. 1978 Critical ionization states in the reaction catalysed by triosephosphate isomerase. *Biochemistry* **17**, 2971–2978.

Belasco, J. G. & Knowles, J. R. 1980 Direct observation of substrate distortion by triosephosphate isomerase using Fourier transform infrared spectroscopy. *Biochemistry* **19**, 472–477.

Blacklow, S. C. & Knowles, J. R. 1990 How can a catalytic lesion be offset? The energetics of two pseudorevertant triosephosphate isomerases. *Biochemistry* **29**, 4099–4108.

Browne, C. A., Campbell, I. D., Kiener, P. A., Phillips, D. C., Waley, S. G. & Wilson, I. A. 1976 Studies of the histidine residues of triosephosphate isomerase by proton magnetic resonance and X-ray crystallography. *J. molec. Biol.* **100**, 319–343.

Bruice, T. C. & Schmir, G. L. 1958 Imidazole catalysis. The reaction of substituted imidazoles with phenyl acetates in aqueous solution. *J. Am. chem. Soc.* **80**, 148–156.

Campbell, I. D., Jones, R. B., Kiener, P. A., Richards, E., Waley, S. G. & Wolfenden, R. 1978 The form of 2-phosphoglycolic acid bound by triosephosphate isomerase. *Biochem. biophys. Res. Commun.* **83**, 347–352.

Collins, K. D. 1974 An activated intermediate analogue. *J. biol. Chem.* **249**, 136–142.

Corey, E. J. & Sneen, R. A. 1956 Stereoelectronic control in enolization-ketonization reactions. *J. Am. chem. Soc.* **78**, 6269–6278.

Cotton, F. A., Hazen, E. E. & Legg, M. J. 1979 *Staphylococcal* nuclease: proposed mechanism of action based on structure of enzyme-thymidine 3′,5′-bis-phosphate-calcium ion complex at 1.5 Å resolution. *Proc. natn. Acad. Sci. U.S.A.* **76**, 2551–2555.

Davenport, R. C. 1986 Studies on the catalytic mechanism of triosephosphate isomerase. Ph.D. Thesis, Massachusetts Institute of Technology, Cambridge, Massachusetts.

de la Mare, S., Coulson, A. F. W., Knowles, J. R., Priddle, J. D. & Offord, R. E. 1972 Active-site labelling of triosephosphate isomerase. *Biochem. J.* **129**, 321–331.

Gandour, R. D. 1981 On the importance of orientation in general base catalysis by carboxylate. *Bioorg. Chem.* **10**, 169–176.

Hartman, F. C. 1968 Irreversible inactivation of triosephosphate isomerase by 1-hydroxy-3-iodo-2-propanone phosphate. *Biochem. biophys. Res. Commun.* **33**, 888–894.

Hegarty, G. G. & Jencks, W. P. 1975 Bifunctional catalysis of the enolization of acetone. *J. Am. chem. Soc.* **97**, 7188–7189.

Hol, W. G. J. 1985 The role of the α-helix dipole in protein function and structure. *Prog. Biophys. molec. Biol.* **45**, 149–195.

Joseph, D., Petsko, G. A. & Karplus, M. 1990 Anatomy of a conformational change: hinged 'lid' motion of the triosephosphate isomerase loop. *Science, Wash.* **249**, 1425–1428.

Koshland, D. E. Jr. 1959 Enzyme flexibility and enzyme action. *J. cell. comp. Physiol.* **54**, (Suppl. 1), 245–258.

Komives, E. A., Chang, L. C., Lolis, E., Tilton, R. F., Petsko, G. A. & Knowles, J. R. 1991 Electrophilic catalysis in triosephosphate isomerase: the role of histidine-95. *Biochemistry* (In the press.).

Kuliopulos, A., Mildvan, A. S., Shortle, D. & Talalay, P. 1989 Kinetic and ultraviolet spectroscopic studies of active-site mutants of Δ^5-3-ketosteroid isomerase. *Biochemistry* **28**, 149–159.

Lambeir, A.-M., Opperdoes, F. R. & Wierenga, R. K. 1987 Kinetic properties of triosephosphate isomerase from *Trypanosoma brucei brucei*. *Eur. J. Biochem.* **168**, 69–74.

Lolis, E., Alber, T., Davenport, R. C., Rose, D., Hartman, F. C. & Petsko, G. A. 1990 Structure of yeast triosephosphate isomerase at 1.9 Å resolution. *Biochemistry* **29**, 6609–6618.

Lolis, E. & Petsko, G. A. 1990 Crystallographic analysis of the complex between triosephosphate isomerase and 2-phosphoglycolate at 2.5 Å resolution: implications for catalysis. *Biochemistry* **29**, 6619–6625.

Luecke, H. & Quiocho, F. A. 1990 High specificity of a phosphate transport protein determined by hydrogen bonds. *Nature, Lond.* **347**, 402–406.

Maynard-Smith, J. 1970 Natural selection and the concept of protein space. *Nature, Lond.* **225**, 563–564.

Nickbarg, E. B., Davenport, R. C., Petsko, G. A. & Knowles, J. R. 1988 Triosephosphate isomerase: removal of a putatively electrophilic histidine residue results in a subtle change in catalytic mechanism. *Biochemistry* **27**, 5948–5960.

Pauling, L. 1960 The nature of the chemical bond, 3rd edn. New York: Cornell University Press.

Perutz, M. F., Gronenborn, A. M., Clore, G. M., Fogg, J. H.

& Shih, D. T-b. 1985 The pK_a values of two histidine residues in human haemoglobin, the Bohr effect, and the dipole movements of α-helices. *J. molec. Biol.* **183**, 491–498.

Pompliano, D. L., Peyman, A. & Knowles, J. R. 1990 Stabilization of a reaction intermediate as a catalytic device: definition of the functional role of the flexible loop in triosephosphate isomerase. *Biochemistry* **29**, 3186–3194.

Richard, J. P. 1984 Acid-base catalysis of the elimination and isomerization reactions of triose phosphates. *J. Am. chem. Soc.* **106**, 4926–4936.

Rieder, S. V. & Rose, I. A. 1959 The mechanism of the triosephosphate isomerase reaction. *J. biol. Chem.* **234**, 1007–1010.

Rose, I. A. 1962 Mechanisms of C–H bond cleavage in aldolase and isomerase reactions. *Brookhaven Symp. Biol.* **15**, 293–309.

Sali, D., Bycroft, M. & Fersht, A. R. 1988 Stabilization of protein structure by interaction of α-helix dipole with a charged side chain. *Nature, Lond.* **335**, 740–743.

Sowadski, J. M., Handschumacher, M. D., Krishna Murthy, H. M., Foster, B. A. & Wyckoff, H. W. 1985 Refined structure of alkaline phosphatase from *Escherichia coli* at 2.8 Å resolution. *J. molec. Biol.* **186**, 417–433.

Waley, S. G., Miller, J. C., Rose, I. A. & O'Connell, E. L. 1970 Identification of site in triosephosphate isomerase labelled by glycidol phosphate. *Nature, Lond.* **227**, 181.

Walba, H. & Isensee, R. W. 1961 Acidity constants of some arylimidazoles and their cations. *J. org. Chem.* **26**, 2789–2791.

Webb, M. R. & Knowles, J. R. 1975 The orientation and accessibility of substrates on the active site of triosephosphate isomerase. *Biochemistry* **14**, 4692–4698.

Terpenoid biosynthesis and the stereochemistry of enzyme-catalysed allylic addition–elimination reactions

DAVID E. CANE, CHRISTOPHER ABELL, PAUL H. M. HARRISON, BRIAN R. HUBBARD, CHARLES T. KANE, RENE LATTMAN, JOHN S. OLIVER AND STEVEN W. WEINER

Department of Chemistry, Brown University, Providence, Rhode Island 02912, U.S.A.

SUMMARY

Allylic addition–elimination reactions are widely used in the enzyme-catalysed formation of terpenoid metabolites. It has earlier been shown that the isoprenoid chain elongation reaction catalysed by farnesyl pyrophosphate synthase involving successive condensations of dimethylallyl pyrophosphate (DMAPP) and geranyl pyrophosphate (GPP) with isopentenyl pyrophosphate (IPP) corresponds to such an $S_{E'}$ reaction with net syn stereochemistry for the sequential electrophilic addition and proton elimination steps. Studies of the enzymic cyclization of farnesyl pyrophosphate (FPP) to pentalenene have now established the stereochemical course of two additional biological $S_{E'}$ reactions. Incubation of both $(9R)$- and $(9S)$-[9-^{3}H, 4,8-^{14}C]FPP with pentalenene synthase and analysis of the resulting labelled pentalenene has revealed that H-9*re* of FPP becomes H-8 of pentalenene, while H-9*si* undergoes net intramolecular transfer to the adjacent carbon, becoming H-1*re* (H-1α) of pentalenene, as confirmed by subsequent experiments with [10-^{2}H, 11-^{13}C]FPP. These results correspond to net anti-stereochemistry in the intramolecular allylic addition–elimination reaction. The stereochemical course of a second $S_{E'}$ reaction has now been examined by analogous incubations of $(4S, 8S)$-[4,8-^{3}H, 4,8-^{14}C]FPP and $(4R, 8R)$-[4,8-^{3}H, 4,8-^{14}C]FPP with pentalenene synthase. Determination of the distribution of label in the derived pentalenenes showed stereospecific loss of the original H-8*si* proton. Analysis of the plausible conformation of the presumed reaction intermediates revealed that the stereochemical course of the latter reaction cannot properly be described as either syn or anti, since cyclization and subsequent double bond formation require significant internal motions to allow proper overlap of the scissile C–H bond with the developing carbocation.

INTRODUCTION

Cyclic sesquiterpenes are widely distributed plant and microbial metabolites. Although so far some 200 different sesquiterpene carbon skeletons have been identified, it is believed that all of these are derived from a single acyclic precursor, farnesyl pyrophosphate (FPP) (**1**) by variations of a common cyclization mechanism (Ruzicka *et al.* 1953; Ruzicka 1959; Ruzicka 1963). According to the currently accepted picture, ionization of the allylic pyrophosphate ester is followed by electrophilic attack of the resulting allylic cation on one of the remaining two double bonds of the substrate (figure 1). The resulting cationic intermediates can undergo further electrophilic reactions, including chemically well-precedented rearrangements, with eventual quenching of the positive charge by deprotonation or capture of an external nucleophile, such as water. Support for these schemes has come not only from classical biosynthetic precursor-product experiments with intact organisms (Cane 1981; Croteau 1981), but from detailed studies of a small group of sesquiterpene synthases and related monoterpene cyclases (Cane 1990; Croteau 1987).

One of the most fundamental bond-forming motifs in the biosynthesis of cyclic terpenes is the generation of a new carbon–carbon bond by the addition of a carbenium ion to a double bond, followed by loss of one of the original allylic protons with formation of a new double bond (Cane 1980) (figure 2). In principle, this transformation can occur with net suprafacial or antarafacial stereochemistry, depending on whether the allylic C–H bond undergoing cleavage is syn or anti to the face of the double bond undergoing electrophilic attack. The prototype of this electrophilic allylic addition–elimination sequence, formally an $S_{E'}$ transformation, is the terpenoid chain-elongation reaction carried out by the family enzymes known as prenyl transferases (Poulter & Rilling 1981) (figure 3). For example, farnesyl pyrophosphate synthase catalyses the condensation of dimethylallyl pyrophosphate (DMAPP) (**2**) with two equivalents of isopentenyl pyrophosphate (IPP) (**3**). Ionization of DMAPP yields an allylic cation which forms a new C–C bond by electrophilic attack on C-4 of the co-substrate IPP. Subsequent loss of one of the C-2 protons generates the intermediate allylic pyrophosphate, geranyl pyrophosphate (GPP) (**4**), which itself undergoes an

Phil. Trans. R. Soc. Lond. B (1991) **332**, 123–129

Printed in Great Britain

[17]

123

9-2

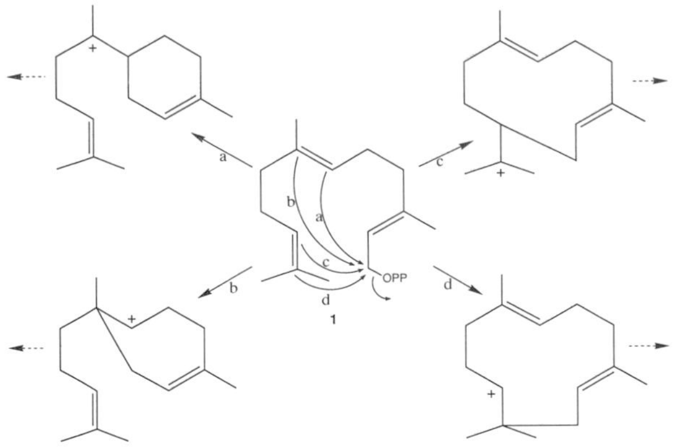

Figure 1. Cyclization of farnesyl pyrophosphate to cyclic sesquiterpenes.

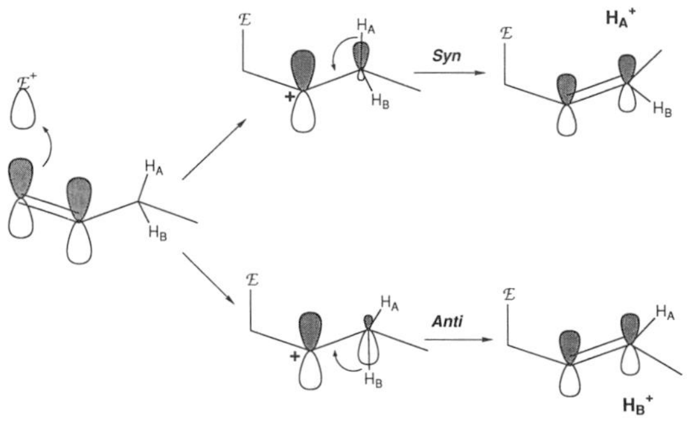

Figure 2. Syn and anti electrophilic allylic addition–elimination ($S_{E'}$) reactions.

analogous condensation–elimination with IPP to yield farnesyl pyrophosphate (**1**). Analogous chain-elongation reactions account for the formation of polyisoprenoids ranging in length from four to several thousand isoprenoid units.

The stereochemical course of the farnesyl pyrophosphate synthase reaction was established nearly 25 years ago by Cornforth & Popjak as part of their landmark studies of cholesterol biosynthesis (Donninger & Popjak 1966; Cornforth *et al.* 1966 *a*, *b*). In an elegant series of experiments, these investigators showed that displacement of the pyrophosphate moiety from C-1 of both allylic substrates, DMAPP and GPP, takes place with net inversion of configuration (figure

3). They also showed that electrophilic attack by each allylic cation takes place exclusively on the *re*-face of the IPP double bond coupled with stereospecific loss of the 2*re* proton (H$_D$ in figure 3), corresponding to net syn or suprafacial stereochemistry for the $S_{E'}$ reaction. In a related series of experiments, it was also found that formation of the *cis* double bonds of the polyisoprenoid rubber involved loss of the 2*si* proton of IPP. These results were interpreted as reflecting an alternative folding of the IPP substrate at the active site of the prenyl transferase with preservation of the overall syn stereochemistry of the allylic addition–elimination reaction. Remarkably, recent studies have shown that the inherent prenyl transferase activity of *Hevea* latex is indistinguishable from farnesyl pyrophosphate synthase in chain length specificity and stereochemical preference (Light *et al.* 1989; Dennis & Light 1989; Dennis *et al.* 1989). Instead, the ability of the rubber prenyl transferase to mediate the formation of polymeric *cis*-isoprenoids is modulated by a low molecular mass protein designated as rubber elongation factor.

The prenyl chain-elongation reaction is the intermolecular counterpart of the intramolecular cyclization reactions catalysed by sesquiterpene and monoterpene synthases. For example, the first step in the enzymic conversion of farnesyl pyrophosphate to aristolochene (**5**) is believed to be an electrophilic attack at C-10 of the distal double bond followed by loss of a proton from the C-12 (*cis*) methyl group of FPP to generate the enzyme-bound intermediate, germacrene A (**6**) (Cane *et al.* 1989 *b*; Cane *et al.* 1990 *a*) figure 4). Similarly, the last step in the conversion of farnesyl pyrophosphate to β-*trans*-bergamotene (**7**) is thought to be cyclization of the derived bisabolyl cation **8** by attack on the cyclohexene double bond and loss of a proton from the attached methyl group (Cane *et al.* 1989 *a*) (figure 5).

For the past few years we have been studying the enzymic cyclization of farnesyl pyrophosphate to the tricyclic sesquiterpene pentalenene (**9**), the parent hydrocarbon of the pentalenolactone family of antibiotics (Cane & Tillman 1983; Cane *et al.* 1984; Cane *et al.* 1990 *b*). This intriguing transformation, which is catalysed by a soluble, monomeric enzyme of M_r 43 ± 1.5 kilodaltons (kD) isolated from *Streptomyces*

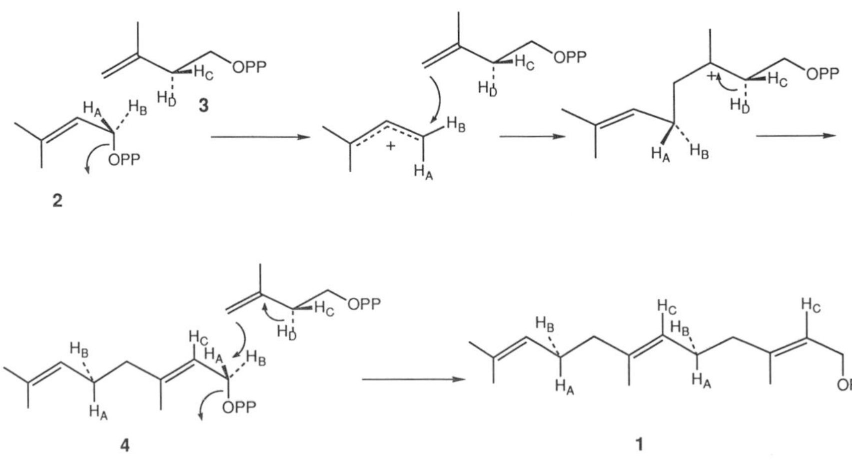

Figure 3. Formation of farnesyl pyrophosphate (**1**) by stepwise condensation of DMAPP (**2**) with IPP (**3**) catalysed by FPP synthase.

Phil. Trans. R. Soc. Lond. B (1991)

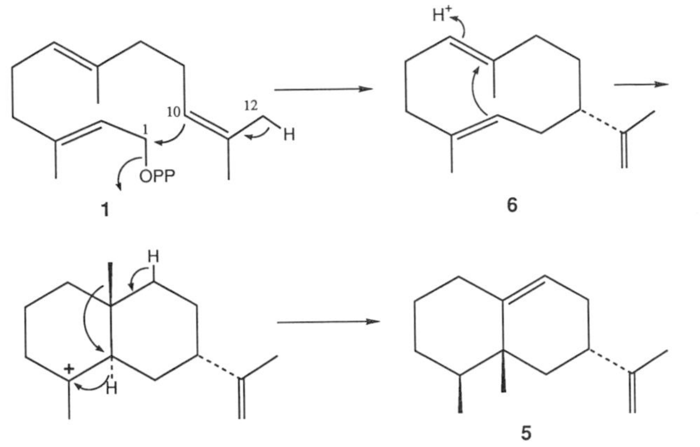

Figure 4. Enzymic conversion of FPP (**1**) to aristolochene (**5**), illustrating $S_{E'}$ reaction in the cyclization of FPP to germacrene A (**6**).

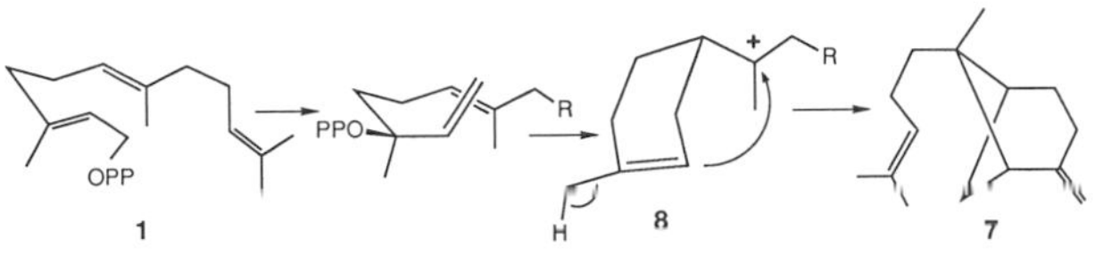

Figure 5. Enzymic conversion of FPP (**1**) to β-*trans*-bergamotene (**7**) involving $S_{E'}$ cyclization of bisabolyl cation **8** to **7**.

UC5319, involves two formal $S_{E'}$ reactions (figure 6). In the first, ionization of FPP and electrophilic attack of the resulting cation on C-11 of the distal double bond is thought to generate the humulyl cation **10** which undergoes deprotonation to form the intermediate humulene (**11**). Following reprotonation and further cyclization, the derived protoilludyl cation **12** can undergo a hydride shift, setting the stage for a second $S_{E'}$ reaction. In the course of the latter transformation, transannular attack on the remaining *trans* double bond in **13** and subsequent deprotonation yields pentalenene.

Incubation of [9-³H]FPP (**1**, $H_A=H_B=T$) with pentalenene synthase and determination of the distribution of isotopic label in the derived pentalenene had shown that the product retained the bulk of the original tritium, with half at the expected position C-

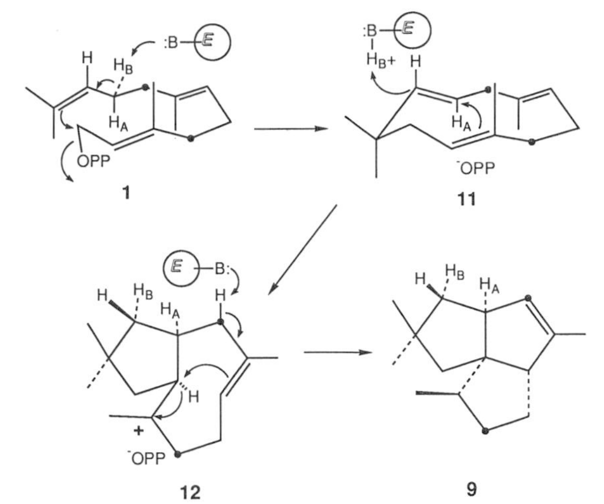

Figure 7. Enzymic cyclization of (9R)- and (9S)-[9-³H, 4,8-¹⁴C]FPP (**1**, $H_A=T, H_B=H$; and **1**, $H_A=H, H_B=T$) to pentalenene (**9**).

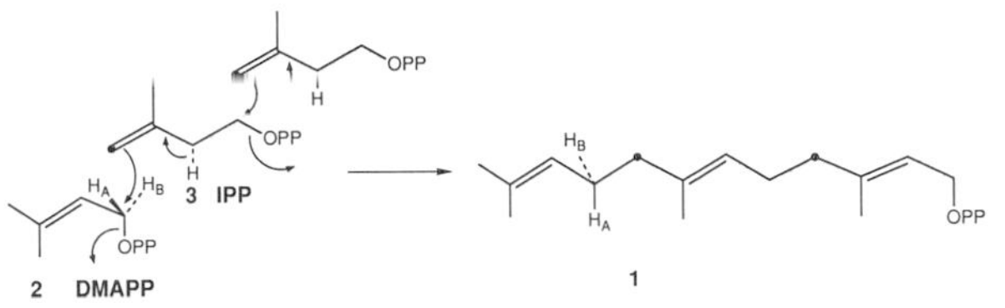

Figure 8. Enzymic synthesis of (9R)- and (9S)-[9-³H, 4,8-¹⁴C]FPP (**1**, $H_A=T, H_B=H$; and **1**, $H_A=H, H_B=T$) from (1R)- and (1S)-[1-³H]DMAPP (**2**, $H_A=T, H_B=H$; and **2**, $H_A=H, H_B=T$) and [4-¹⁴C]IPP (**3**).

8, the remainder having been transferred to C-1 (Cane *et al.* 1984) (figure 7). Although formation of humulene requires loss of one of the original C-9 protons of FPP, this same proton is returned in the subsequent deprotonation step, without apparent exchange with the medium. To determine the stereochemical course of the allylic addition–elimination reaction, we required samples of FPP stereospecifically tritiated at C-9. The individual samples of (9R)- and (9S)-[9-³H, 4,8-¹⁴C]FPP (**1**, $H_A=T, H_B=H$; and **1**, $H_A=H, H_B=T$) could be prepared by using avian prenyl transferase to catalyse the condensation of (1R)- and (1S)-[1-³H]DMAPP (**2**, $H_A=T, H_B=H$; and **2**,

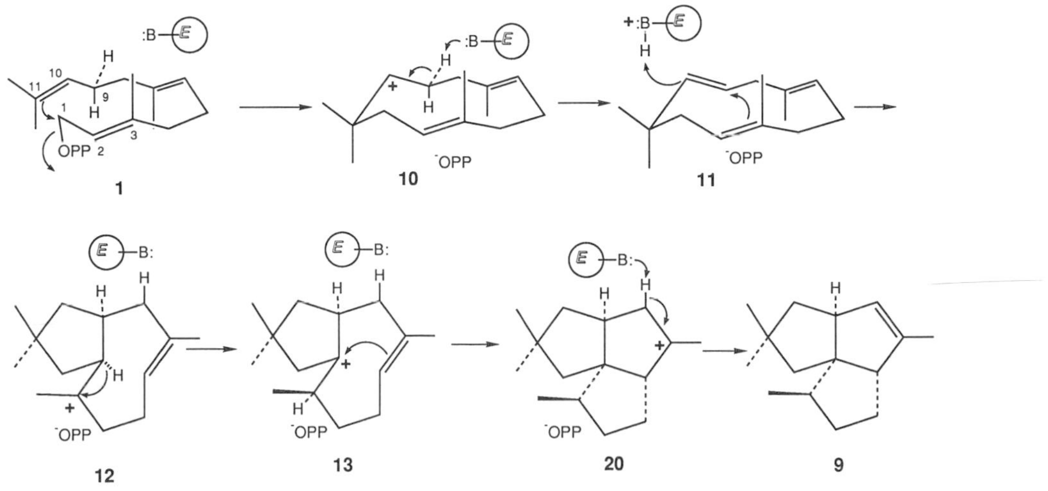

Figure 6. Enzymic cyclization of FPP (**1**) to pentalenene (**9**). The conversion of **1** to humulene (**11**).

[19]

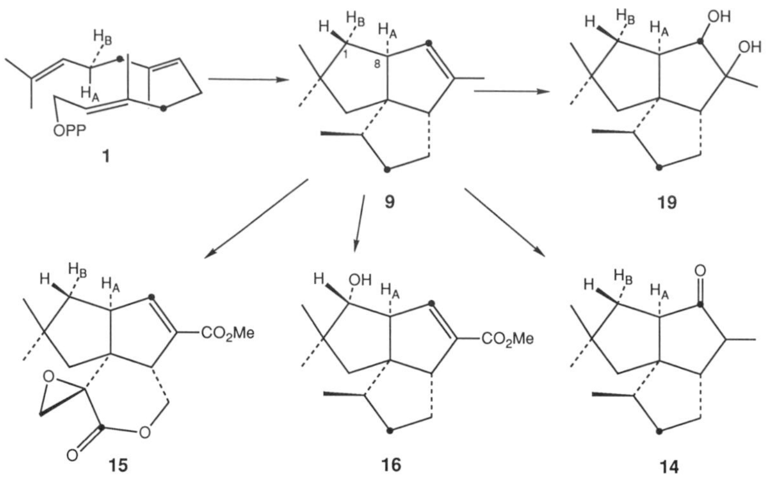

Figure 9. Chemical and microbial degradation of pentalenene (**9**) derived from (9R)- and (9S)-[9-³H, 4,8-¹⁴C]FPP (**I**, H_A=T, H_B=H; and **I**, H_A=H, H^B=T).

H_A=H, H_B=T), respectively, with [4-¹⁴C]IPP (**3**) (figure 8). Incubation of the resulting stereospecifically labelled FPP samples with pentalenene synthase gave labelled pentalenene, which was analysed by a combination of chemical and microbial degradation methods (Cane *et al.* 1990 *b*) (figure 9). Thus treatment of pentalenene derived from (9R)-[9-³H, 4,8-¹⁴C]FPP with diborane followed by oxidation with PCC gave the ketone **14** (H_A=T, H_B=H) which lost all tritium upon base catalysed exchange. In a complementary series of experiments, the tritium corresponding to H-9*si* of FPP was located at C-1 of pentalenene by refeeding labelled pentalenene derived from (9S)-[9-³H, 4,8-¹⁴C]FPP to growing cultures of *Streptomyces* UC5319. Whereas the ³H:¹⁴C ratio of the resulting epipentalenolactone F methyl ester (**15**, H_A=H, H_B=T) was unchanged with respect to the original FPP and derived pentalenene, the corresponding sample of pentalenic acid methyl ester (**16**) was devoid of tritium. Since earlier experiments involving [¹³C]-labelled substrates had already established that the initial electrophilic attack takes place on the si face of the 10,11-double bond of FPP, the allylic addition–elimination must take place with net anti-stereochemistry, in contrast to the demonstrated syn stereochemistry of the prototype prenyl transferase reaction. The observed stereochemistry of the pentalenene synthase-catalysed reaction suggests that the folding of the substrate prevents access of the enzymic base to the H-9*re* proton (figure 7).

In a related series of experiments we have also determined the stereochemical course of the reprotonation reaction in which the original H-9*si* proton of FPP undergoes net intramolecular transfer of H-1 of pentalenene (Cane *et al.* 1990 *b*) (figure 10). Thus incubation of [10-²H, 11-¹³C]FPP with pentalenene synthase and analysis of the derived labelled pentalenene by a combination of ¹³C and ²H nuclear magnetic resonance (NMR) spectroscopy established that the deuterium atom from C-10 of FPP occupied exclusively the H-1*si* (H-1β) position in **9**, indicating that protonation occurred on the 10 *re* face of the C-9,10 double bond of the intermediate humulene.

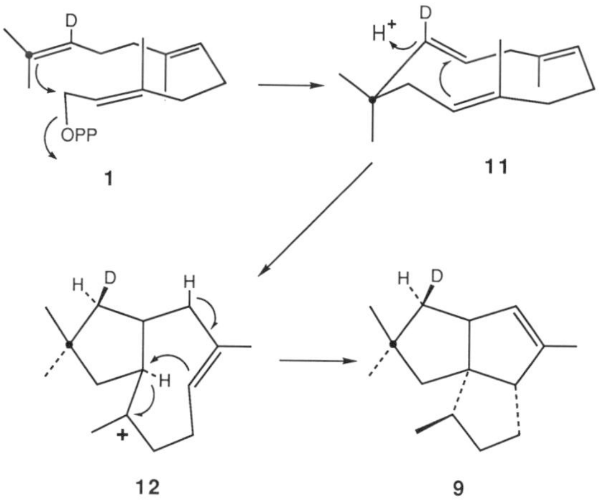

Figure 10. Cyclization of [10-²H, 11-¹³C]FPP (**I**) to pentalenene (**9**).

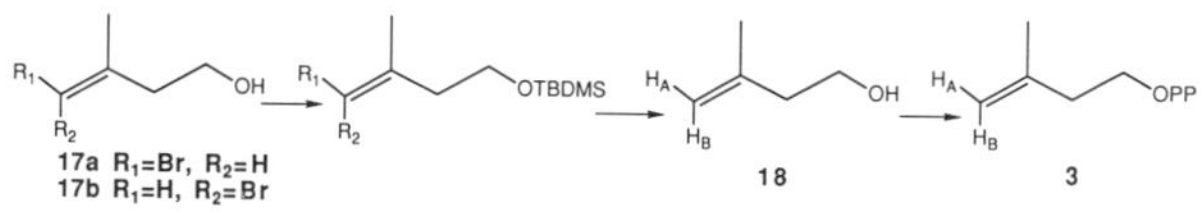

Figure 11. Preparation of (E)- and (X)-[4-³H]isopentenyl pyrophosphate (**3**, H_A=T, H_B=H; and **3**, H_A=H, H_B=T) from (E)- and (X)-bromoisopentenols (**17a** and **17b**).

We have earlier shown that the formation of pentalenene involves loss of one of the two hydrogen atoms originally attached to C-8 of farnesyl pyrophosphate, presumably in the final step of the overall cyclization reaction (Cane & Tillman 1983). To establish the stereochemical course of this second $S_{E'}$ process, we turned once again to the prenyl transferase reaction as a means of preparing the requisite samples of FPP, in this case stereospecifically labelled with tritium at C-8. To this end, (E)- and (Z)-bromoisopentenols (**17a** and **17b**), prepared by literature methods (Cornforth *et al.* 1966 *b*), were separately protected as the *t*-butyldimethylsilyl (TBDMS) ethers and metallated by treatment with *t*-butyllithium (Ito *et al.* 1987; Cane *et al.* 1990 *a*) (figure 11). The resulting vinyllithium intermediates were each quenched with

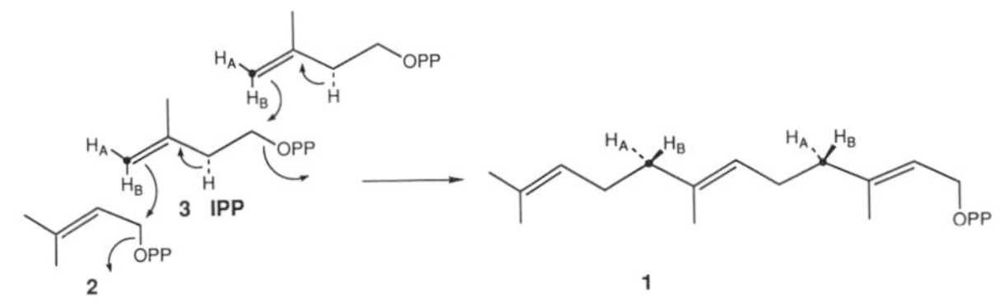

Figure 12. Enzymic synthesis of $(4S, 8S)$-[4,8-^{3}H, 4,8-^{14}C]FPP (**I**, H_A=T, H_B=H) and $(4R,8R)$-[4,8-^{3}H, 4,8-^{14}C]FPP (**I**, H_A=H, H_B=T) from stereospecifically tritiated IPP (**3**).

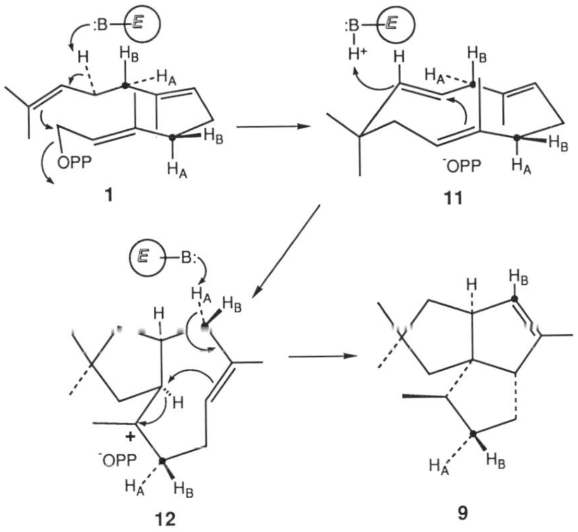

Figure 13. Cyclization of $(4S,8S)$-[4,8-^{3}H, 4,8-^{14}C]FPP (**I**, H_A=T, H_B=H) and $(4R, 8R)$-[4,8-^{3}H, 4,8-^{14}C]FPP (**I**, H_A=H, H_B=T) to pentalenene (**9**).

tritiated trifluoroacetic and deprotected to yield the individual (E)- and (Z)-[4-^{3}H]isopentenols (**18**, H_A=T, H_B=H; and **18**, H_A=H, H_B=T). NMR analysis of the corresponding deuterated isopentenols confirmed the essentially complete stereospecificity of the isotopic labelling procedure. Tosylation and displacement with tris(tetrabutylammonium) pyrophosphate gave (E)- and (Z)-[4-^{3}H]isopentenyl pyrophosphate (**3**, H_A=T, H_B=H; and **3**, H_A=H, H_B=T), respectively (Davisson *et al.* 1986). The individual tritiated IPP samples, mixed with [4-^{14}C]IPP were then incubated with DMAPP and avian prenyl transferase to generate the desired samples of $(4S,8S)$-[4,8-^{3}H,4,8-^{14}C]FPP (**I**, H_A=T, H_B=H) and $(4R,8R)$-[4,8-^{3}H, 4,8-^{14}C]FPP (**I**, H_A=H, H_B=T) (figure 12). Incubation of $(4S,8S)$-[4,8-^{3}H, 4,8-^{14}C]FPP with pentalenene synthase gave pentalenene (**9**, H_A=T, H_B=H), which had lost half of the original tritium, based on comparison of the ^{3}H:^{14}C ratio of the derived crystalline diols **19** with the corresponding isotope ratio of the diphenylurethane derivative of the original farnesol (figure 14). Hydroboration–oxidation of **9** in the manner described above gave the derived ketone **14** (H_A=T, H_B=H) without further loss of tritium. In the complementary series of experiments, cyclization of $(4R,8R)$-[4,8-^{3}H, 4,8-^{14}C]FPP gave pentalenene of unchanged ^{3}H:^{14}C ratio, which upon conversion to the ketone **14** (H_A=H, H_B=T) lost the expected one equivalent of tritium. These results establish that the second allylic addition–elimination

reaction catalysed by pentalenene synthase involves exclusive loss of the H-8*si* proton of farnesyl pyrophosphate.

Intriguingly, the latter $S_{E'}$ transformation involves neither simple syn nor anti stereochemistry, but rather initial electrophilic attack *orthogonal* to the C–H bond to be broken (figure 15). Based on the absolute configuration of the eventually formed pentalenene, it is clear that cyclization of the bicyclic protoilludane intermediate **13** involves electrophilic attack on the *si* (inner) face of the *trans*-6,7-double bond. Considering the presumed conformation of the latter intermediate, it is evident that the p-orbital at C-7 is essentially staggered with respect to the adjacent pair of allylic protons. As the new C–C bond is formed, the attached methyl group at the now positively charged C-7 carbon atom must rotate outwards. Remarkably, although a rotation of approximately only 30° would suffice for proper overlap of the vacant p-orbital with the adjacent β-proton (originally H-8*re* of FPP), the demonstrated loss of H-8*si* of FPP implies a rotation of some 90°. Interestingly, the latter rotation brings the lobe of the p-orbital originally on the *si* face of the 6,7-double bond (i.e. syn to the newly formed C–C bond) into overlap with the C–H bond undergoing cleavage. In the latter sense, therefore, the allylic addition–elimination sequence may be considered to take place with net syn stereochemistry, although in the most likely reactive conformation, both protons are originally anti to the attacking electrophile!

Taken together, the above results show that there is no stereochemical imperative governing biochemical $S_{E'}$ reactions. The overall stereochemistry of such electrophilic allylic addition–elimination reactions is most likely the result of the relative arrangement at the active site of the electrophile and the base which eventually quenches the positive charge. For example, IPP isomerase catalyses the prototropic rearrangement of IPP to DMAPP (figure 16). In contrast to the prenyl transferase reaction, protonation at C-4 and loss of the H-2*re* (H_A) proton of IPP have been shown to take place with net anti stereochemistry (Poulter & Rilling 1981). Interestingly, in recent experiments with cloned IPP isomerase, Poulter has reported that equilibration of IPP and DMAPP in deuterated buffer results in slow exchange not only of H-2*si* (H_B) but of all six allylic methyl protons as well (Street *et al.* 1990). It is proposed that the overall stereochemical course of the allylic addition–elimination reaction is invariant, but that the observed scrambling results from binding of different conformations of the substrate at the active site of the isomerase. Significantly, this relaxation in binding specificity would have no observable consequence on product structure or geometry, and would be unrecognizable in the absence of isotopic labelling. Poulter, based on an earlier suggestion of Popjak, has also proposed that, for the prenyl transferase-catalysed condensation reaction, the base is in fact the inorganic pyrophosphate counter ion which is released in the initial ionization step, thereby accounting for the net syn stereochemistry of isoprenoid chain elongation (Poulter & Rilling 1978). Indeed, consistent with this hypothesis, Poulter has shown that incubation of

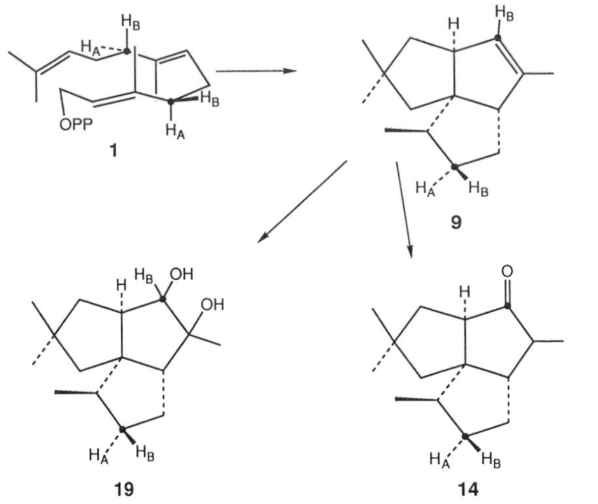

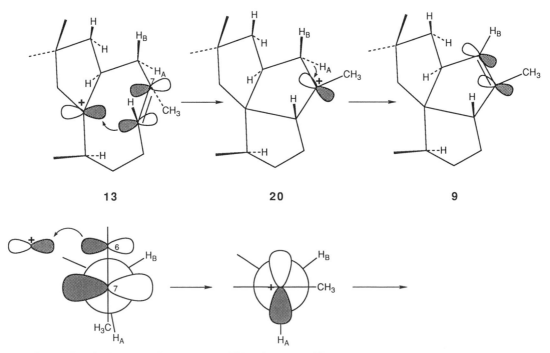

Figure 14. Determination of the distribution of tritium in pentalenene (**9**) derived from $(4S, 8S)$-$[4,8$-$^3H, 4,8$-$^{14}C]$FPP (**1**, H_A=T, H_B=H) and $(4R, 8R)$-$[4,8$-$^3H, 4,8$-$^{14}C]$FPP (**1**, H_A=H, H_B=T).

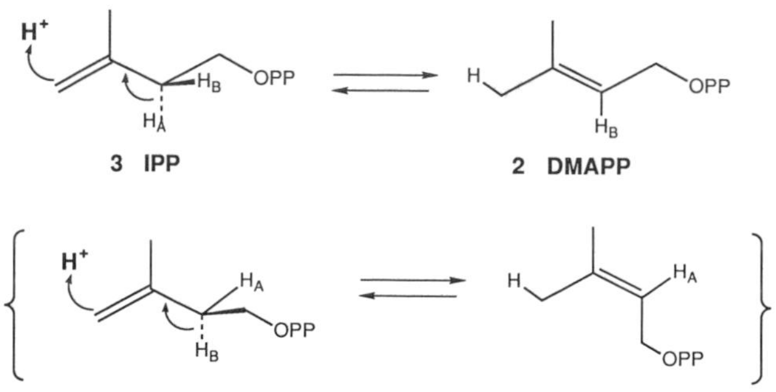

Figure 16. Isomerase catalysed interconversion of IPP and DMAPP by allylic proton addition–elimination with net anti-stereochemistry. The minor reaction leading to the same products but by anomalous binding of substrate is also shown.

suitably designed bisubstrate analogs with FPP synthase results in formation of cyclic products (Davisson *et al.* 1985).

For the pentalenene synthase reaction, we have proposed that a single base is responsible for the sequential deprotonation–reprotonation–deprotonation events taking place at C-9,-10, and -8 of the farnesyl pyrophosphate substrate (figure 13). The above-described elucidation of the stereochemical course of the individual proton transfer processes is consistent with, but does not require, the action of a single enzymic base. For example, in removal of H-9α from the pentalenyl cation **20**, the C–H bond being broken is roughly parallel to the C-1–H-1α bond which has

been generated in the earlier reprotonation step. Identification of the actual active site base should become possible once cloned pentalenene synthase is available.

This work was supported by a grant no. GM22172 from the U.S. National Institutes of Health.

REFERENCES

Cane, D. E. 1980 The stereochemistry of allylic pyrophosphate metabolism. *Tetrahedron* **36**, 1109.

Cane, D. E. 1981 The biosynthesis of sesquiterpenes. In *Biosynthesis of isoprenoid compounds* (ed. J. W. Porter & S. L. Spurgeon), vol. 1, pp. 283–374. New York: J. W. Wiley & Sons.

Cane, D. E. 1990 The enzymatic formation of sesquiterpenes. *Chem. Rev.* **9**, 1089.

Cane, D. E. & Tillman, A. M. 1983 Pentalenene biosynthesis and the enzymatic cyclization of farnesyl pyrophosphate. *J. Am. chem. Soc.* **105**, 122.

Figure 15. Stereochemical course of conversion of bicyclic cation **13** to pentalenene illustrating orthogonal relation between the direction of the initial electrophilic attack on the 6,7-double bond and the eventually broken C–H_A bond. The two Newman projections are along the 7,8 bond of **13** and **20**, respectively.

Cane, D. E., Abell, C. & Tillman, A. M. 1984 Pentalenene biosynthesis and the enzymatic cyclization of farnesyl pyrophosphate. Proof that the cyclization is catalysed by a single enzyme. *Bioorg. Chem.* **12**, 312.

Cane, D. E., McIlwaine, D. B. & Harrison, P. H. M. 1989*a* Bergamotene biosynthesis and the enzymatic cyclization of farnesyl pyrophosphate. *J. Am. chem. Soc.* **111**, 1152.

Cane, D. E., Prabhakaran, P. C., Salaski, E. J., Harrison, P. H. M., Noguchi, H. & Rawlings, B. J. 1989*b* Aristolochene biosynthesis and enzymatic cyclization of farnesyl pyrophosphate. *J. Am. chem. Soc.* **111**, 8914.

Cane, D. E., Prabhakaran, P. C., Oliver, J. S. & McIlwaine, D. B. 1990*a* Aristolochene biosynthesis. Stereochemistry of the deprotonation steps in the enzymatic cyclization of farnesyl pyrophosphate. *J. Am. chem. Soc.* **112**, 3209.

Cane, D. E., Oliver, J. S., Harrison, P. H. M., Abell, C., Hubbard B. R., Kane, C. T. & Lattman, R. 1990*b* The biosynthesis of pentalenene and pentalenolactone. *J. Am. chem. Soc.* **112**, 4513.

Cornforth, J. W., Cornforth, R. H., Donninger, C. & Popjak, G. 1966*a* Studies on the Biosynthesis of Cholesterol. XIX. Steric Course of Hydrogen Eliminations and of C–C Bond Formations in Squalene Biosynthesis. *Proc. R. Soc. Lond.* B **163**, 492.

Cornforth, J. W., Cornforth, R. H., Popjak, G. & Yengoyan, L. 1966*b* Studies on the Biosynthesis of Cholesterol. XX. Steric Course of Decarboxylation of 5-Pyrophospho-mevalonate and of the Carbon to Carbon Bond Formation in the Biosynthesis of Farnesyl Pyrophosphate. *J. biol. Chem.* **241**, 3970.

Croteau, R. 1981 Biosynthesis of monoterpenes. In *Biosynthesis of isoprenoid compounds* (ed. J. W. Porter & S. L. Spurgeon), vol. 1, pp. 225–282. New York: Wiley.

Croteau, R. 1987 Biosynthesis and catabolism of mono-terpenoids. *Chem. Rev.* **87**, 929.

Davisson, V. J., Neal, T. R. & Poulter, C. D. 1985 Farnesylpyrophosphate synthetase. A case for common electrophilic mechanisms for prenyltransferases and terpene cyclases. *J. Am. chem. Soc.* **107**, 5277.

Davisson, V. J., Woodside, A. B., Neal, T. R., Stremler, K. E., Muehlbacher, M. & Poulter, C. D. 1986 Phosphorylation of isoprenoid alcohols. *J. org. Chem.* **51**, 4768.

Dennis, M. S. & Light, D. R. 1989 Rubber elongation factor from *Hevea brasiliensis*. *J. biol. Chem.* **264**, 18608.

Dennis, M. S., Henzel, W. J., Bell, J., Kohr, W. & Light, D. R. 1989 Amino acid sequence of rubber elongation factor protein associated with rubber particles in *Hevea* latex. *J. biol. Chem.* **264**, 18618.

Donninger, C. & Popjak, G. 1966 Studies on the bio-synthesis of cholesterol. XVIII. The stereospecificity of mevaldate reductase and the biosynthesis of asymmetri-cally labelled farnesyl pyrophosphate. *Proc. R. Soc. Lond.* B **163**, 465.

Ito, M., Kobayashi, M., Koyama, T. & Ogura, K. 1987 Stereochemical analysis of prenyltransferase reactions leading to (Z)- and (E)-polyprenyl chains. *Biochemistry* **26**, 4745.

Light, D. R., Lazarus, R. A. & Dennis, M. S. 1989 Rubber elongation by farnesyl pyrophosphate synthases involves a novel switch in enzyme stereospecificity. *J. biol. Chem.* **264**, 18598.

Poulter, C. D. & Rilling, H. C. 1978 The Prenyl Transfer reaction. Enzymatic and mechanistic studies of the 1′-4 coupling reaction in the terpene biosynthetic pathway. *Acc. chem. Res.* **11**, 307.

Poulter, C. D. & Rilling, H. C. 1981 Prenyl transferases and isomerase. In *Biosynthesis of isoprenoid compounds* (ed. J. W. Porter & S. L. Spurgeon), vol. 1, pp. 161–224. New York: Wiley.

Ruzicka, L., Eschenmoser, A. & Heusser, H. 1953 The isoprene rule and the biogenesis of terpenic compounds. *Experientia* **9**, 357.

Ruzicka, L. 1959 History of the isoprene rule. *Proc. Chem. Soc.*, 341.

Ruzicka, L. 1963 Perspektiven der biogenese und der chemie der terpene. *Pure appl. Chem.* **6**, 493.

Street, I. P., Christensen, D. J. & Poulter, C. D. 1990 Hydrogen exchange during the enzyme-catalysed isomer-ization of isopentenyl diphosphate and dimethylallyl diphosphate. *J. Am. chem. Soc.* **112**, 8577.

A structural and mechanistic comparison of pyridoxal 5'-phosphate dependent decarboxylase and transaminase enzymes†

DAVID GANI

Chemistry Department, The Purdie Building, The University, St Andrews, Fife KY16 9ST, U.K.

SUMMARY

Stereochemical studies of three pyridoxal phosphate dependent decarboxylases and serine hydroxymethyltransferase have allowed the dispositions of conjugate acids that operate at the C^α and C-4' positions of intermediate quinoids to be determined. Kinetic work with the decarboxylase group has determined that two different acids are involved, a monoprotic acid and a polyprotic acid. The use of solvent kinetic isotope effects allowed the resolution of chemical steps in the reaction coordinate profile for decarboxylation and abortive transamination and pH-sensitivities gave the molecular pK_a of the monoprotic base. Thus the ε-ammonium group of the internal aldimine-forming lysine residue operates at C-4'-*si*-face of the coenzyme and the imidazolium side chain of an active site histidine residue protonates at C^α from the 4'-*si*-face. Histidine serves two other functions, as a base in generating nitrogen nucleophiles during both transaldimination processes and as a binding group for the α-carboxyl group of substrates. The latter role for histidine was determined by comparison of the sequences for decarboxylase active site tetrapeptides (e.g. —S—X—H—K—) with that for aspartate aminotransferase (e.g. —S—X—A—K—) where it was known, from X-ray studies, that the serine and lysine residues interact with the coenzyme. By using the Dunathan Postulate, the conformation of the external aldimine was modified, and without changing the tetrapeptide conformation, the alanine residue was altered to a histidine. This model for the active site of a pyridoxal dependent decarboxylase was consistent with all available stereochemical and mechanistic data. A similar model for serine hydroxymethyltransferase suggested that previous reports of stereochemical infidelity with decarboxylation substrates were incorrect. A series of careful experiments confirmed this. Hence, no actual examples of non-stereospecific α-amino acid decarboxylation by pyridoxal enzymes exist.

1. INTRODUCTION

Pyridoxal 5'-phosphate is a coenzyme for a vast number of important enzyme-catalysed transformations in amino acid metabolism. Although many PLP-dependent enzymes are known which catalyse chemistry at the β- and γ-carbon atoms of amino acid substrates, our discussion here concentrates on reactions at C^α. The four most common of these are transamination, racemization, decarboxylation and retroaldol cleavage. Because racemization is a rather special reaction, in which active site bases and conjugate acids must act upon both faces of the coenzyme, these reactions will not be considered here. The remaining three types of reaction; transamination, decarboxylation and retroaldol cleavage (scheme 1) will now be considered in detail. Our aims are to learn how the apoenzyme controls the chemistry of the coenzyme such that specific bonds connected to C^α are broken.

In 1966 Dunathan proposed, on the basis of stereoelectronic arguments, that the bonds to be broken

at C^α of PLP-amino acid Schiff's bases should be held at 90° to the plane of the extended conjugated system (Dunathan 1966), see scheme 1. In the early 1980s when the first X-ray crystal structure for a PLP-dependent enzyme, aspartate aminotransferase (AAT), was published it was evident that Dunathan's ideas were correct, at least for AAT. However, no X-ray crystal structures had been reported for a decarboxylase or for serine hydroxymethyltransferase (SHMT). Because the prospects for obtaining structures seemed poor, we decided to use stereochemical and kinetic techniques to discover how these other enzymes might control the chemistry of the cofactor. Before embarking on a detailed analysis of the chemistry catalysed by decarboxylases and SHMT, it is useful to summarize the important features of AAT catalysis.

2. ASPARTATE AMINOTRANSFERASE

Transaminases are the best understood PLP-dependent enzymes. Much of the early work in the area was concerned with assessing the stereochemical course of the reaction with respect to C^α of the substrate and C-4' of the coenzyme. However, the most significant contributions came from more recent X-ray crystal

---

† The earlier parts of this work were done in the Chemistry Department at Southampton University.

Phil. Trans. R. Soc. Lond. B (1991) **332**, 131–139

Printed in Great Britain

[25]

131

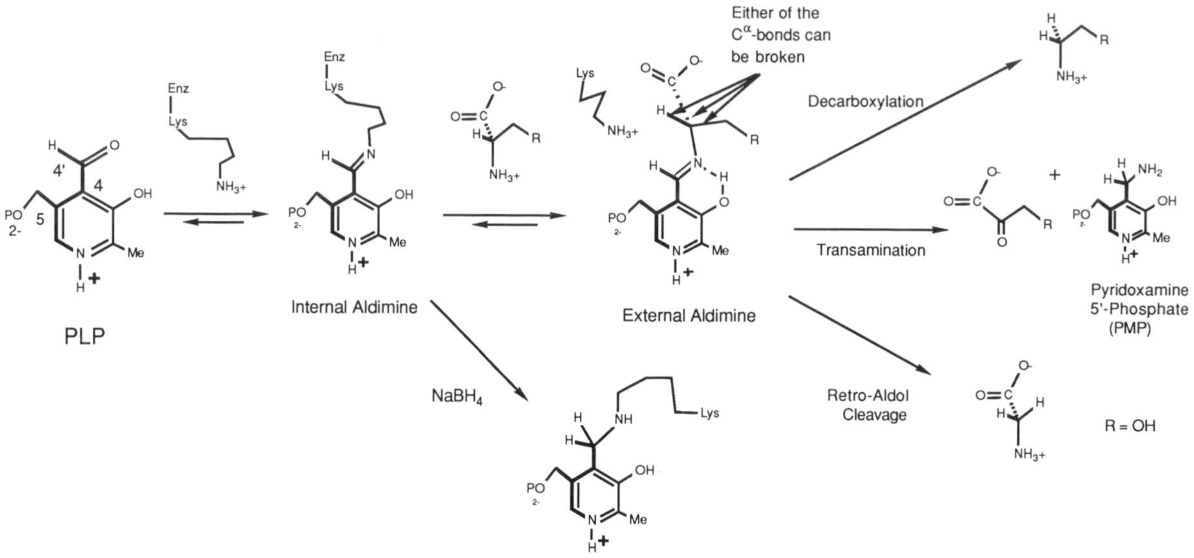

Scheme 1.

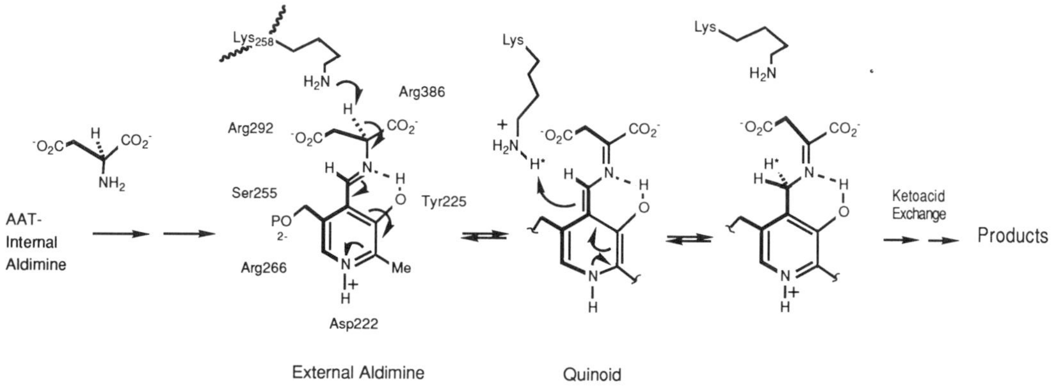

Scheme 2.

work on chicken heart mitochondrial (Kirsch *et al.* 1984) and *Escherichia coli* aspartate aminotransferase and now it is possible to envisage the three-dimensional catalytic function of the enzyme for the entire transamination process, see scheme 2.

The protein is composed of two identical subunits (relative molecular mass, $M_r \approx 45\,000$) which consist of two domains. The coenzyme is bound to the larger domain in a pocket near the subunit interface. The proximal and distal carboxylate groups of the dicarboxylic acid substrates and products (namely, aspartate, glutamate, oxaloacetate and α-keto-glutarate) are bound by two arginine residues (386 and 292 from adjacent subunits). Substrate specificity is determined mainly by these binding interactions. The mode of substrate binding not only ensures efficient catalysis but, also causes a bulk movement in the smaller domain which closes the active-site pocket and moves Arg 386 3 Å† closer to the coenzyme. The transaldimination of the ε-amino group of Lys 258 by the substrate, aspartic acid, to form the substrate aldimine, occurs from the *re*-face at C-4′ and causes the coenzyme to tilt by $\approx 30°$. The released ε-amino group then serves as the enzyme-bound proton carrier for

suprafacial 1,3-prototropic shifts that occur on the *si*-face of C-4′. After the formation and hydrolysis of the initial ketimine, the coenzyme tilt relaxes back slightly.

The coenzyme is held in place by several residues. The protonated pyridinium ring is hydrogen bonded to Asp 222. The 2-methyl group is located in a pocket formed by eight amino acid residues and the 3-oxygen atom is hydrogen bonded to the phenolic OH of Tyr 225. X-ray data also show that a *cisoid* ε-lysine aldimine conformer exists in the absence of substrate (i.e. the aldimine N is on the 3-OH side). As the substrate aldimine also exists in a *cisoid* conformation it seems probable that the conformational changes in the tilt angle of the coenzyme accounts for the differential C-4′-face selectivity of borohydride reducing agents (see scheme 1). The internal and external aldimines are reduced at the C-4′-*re*- and C-4′-*si*-faces, respectively (see Gani 1990).

The 5′-phosphate ester group of the coenzyme appears to be bound by seven or eight hydrogen bonds as the dianion. The guanidinium group of Arg 266 forms two of these H-bonds and offsets the double negative charge (scheme 2). Significantly, the side-chain of Ser 255 (only four residues removed from Lys 258) forms another H-bond to the phosphate ester. This serine residue is conserved in all reported

† $1 \text{ Å} = 10^{-10} \text{ m} = 10^{-1} \text{ nm}$.

Table 1. *Schiff's base-forming region of aspartate amino-transferase enzymes*

E. coli	I	V	A	S	S	Y	S	**K**	N	F	G	L	Y
chicken, mitochondrial	V	L	S	Q	S	Y	A	**K**	N	M	G	L	Y
turkey, mitochondrial	V	L	S	Q	S	Y	A	**K**	N	M	G	L	Y
pig, mitochondrial	C	L	C	Q	S	Y	A	**K**	N	M	G	L	Y
rat, mitochondrial	C	L	C	Q	S	Y	A	**K**	N	M	G	L	Y
human, mitochondrial	C	L	C	Q	S	Y	A	**K**	N	M	G	L	Y
chicken, cytosolic	F	C	A	Q	S	F	S	**K**	N	F	G	L	Y
pig, cytosolic	F	C	A	Q	S	F	S	**K**	N	F	G	L	Y

Table 2. *Schiff's base-forming region of PLP-dependent decarboxylase enzymes*

arginine (E. coli)	A	T	H	S	T	H	**K**	L	L	N	A	L	
glutamate (E. coli)	S	I	S	A	S	G	H	**K**	F				
histidine (Morganella morganii)	S	I	G	V	S	G	H	**K**	M	I	G	S	P
lysine (E. coli)	Y	E	T	E	S	T	H	**K**	L	L	A	A	F
lysine (Hafnia alvei)	Y	E	T	Q	S	T	H	**K**	L	L	A	A	F
ornithine (E. coli)	V	H	**K**	Q	Q	A	G	Q					
dopa (Drosophila)	S	F	N	F	N	P	H	**K**	W	M	L	V	N
dopa (pig)	N	F	N	P	H	**K**	W						
glutamate (feline)	S	V	T	W	N	P	H	**K**	M	M	G	V	L
glycine (chicken)	V	S	H	L	N	L	H	**K**	T	F	C	I	P
SHMT (E. coli)	V	V	T	T	T	H	**K**	T	L	A	G	P	
SHMT (rabbit, cytosolic)	V	V	T	T	T	T	H	**K**	T	L	R	G	C
SHMT (rabbit, mitochondrial)	V	V	T	T	T	T	H	**K**	T	L	R		

sequences for aspartate aminotransferase enzymes (table 1; Smith *et al.* 1991) and, indeed, there is almost 100% homology for all regions of the protein which are involved in binding to the substrate or the coenzyme for each of the sequenced proteins.

3. α-AMINO ACID DECARBOXYLASES

(a) *Stereochemical course at* C^α

By comparison to AAT, the amount of stereo-chemical, mechanistic and structural information available for the decarboxylase group was sparse at the start of our studies. However, the active-site peptide sequences (Smith *et al.* 1991) for some decarboxylases were known (table 2) and the stereochemical courses of decarboxylation had been reported for some systems (Gani 1985). From these studies it was evident that an active site lysine residue was involved in the formation of a coenzyme internal aldimine, as for transaminases and that, in general, PLP-dependent decarboxylases catalysed the decarboxylation of L-amino acid substrates with retention of configuration at C^α. Nevertheless, α,ω-*meso*-diaminopimelate decarboxylase from two different species catalysed the decarboxyl-

ation of the D-amino acid centre of α,ω-*meso*-diaminopimelic acid with inversion of configuration at C^α to give L-lysine (Kelland *et al.* 1985) and the decarboxyl-ase activity of SHMT catalysed the decarboxylation of stereospecifically labelled 2-aminomalonic acid non-stereospecifically to give racemic glycine (Palekar *et al.* 1973). Although it was not clear that only one binding mode was available to the substrate in its interaction with SHMT, it was quite clear that the stereochemical course of PLP-dependent amino acid α-decarboxyl-ations with respect to C^α could not be assumed.

(b) *Coenzyme face selectivity*

To compare the active site of decarboxylases and SHMT with AAT, it would, eventually, be necessary to determine upon which face of the coenzyme decarboxylations occurred. This objective poses a difficult problem because decarboxylases and SHMT, unlike transaminases, do not catalyse chemistry at the C-4′ positions of the quinoid intermediates derived from their physiological substrates, which might allow the reaction at C^α to be assigned to a particular face of the coenzyme (see scheme 2). Fortunately, in the presence of substrate analogues both decarboxylases and SHMT can be forced to catalyse abortive transamination reactions. For example, in the presence of D-alanine, SHMT protonates the coenzyme on the C-4′-*si*-face, the same result as for bona fide trans-aminases (Voet *et al.* 1973). While the observation shows that the enzyme can catalyse at 1,3-proton transfer, it cannot be assumed that the same enzyme-bound base or conjugate acid interacts at both sites. The ambiguity arises because of the uncertainty in knowing how the enzyme binds to D-alanine in the light of Palekar's work (1973) given also that the enzyme can bind to L-alanine (Shostack & Schirch 1988). However, it was evident that formation of PMP occurred stereospecifically and from the same face of the coenzyme as for transaminases.

The situation was even more ambiguous for the decarboxylases where the only reported study of the facial selectivity for a decarboxylase, *E. coli* glutamate decarboxylase, used a racemic non-physiological sub-strate, (2RS)-2-methylglutamic acid, to enhance the frequency of events leading to the formation of PMP via abortive decarboxylation–transamination (Suk-hareva & Braunstein 1971). The results showed that a proton was transferred to the C-4′-*si*-face of the coenzyme. Again, it could not be assumed that the same stereochemical course would have been followed in the much less frequent transamination reaction which occurs with the physiological substrate, or that the apparent 1,5- or 1,3-proton shift occurred supra-facially (Yamada & O'Leary 1977).

(c) L-*methionine decarboxylase*

In view of the confusion and uncertainty regarding the stereochemical courses of events which occur at C^α and C-4′ in decarboxylases and SHMT, a system eminently suited for stereochemical and mechanistic study was sought. L-Methionine decarboxylase (MD)

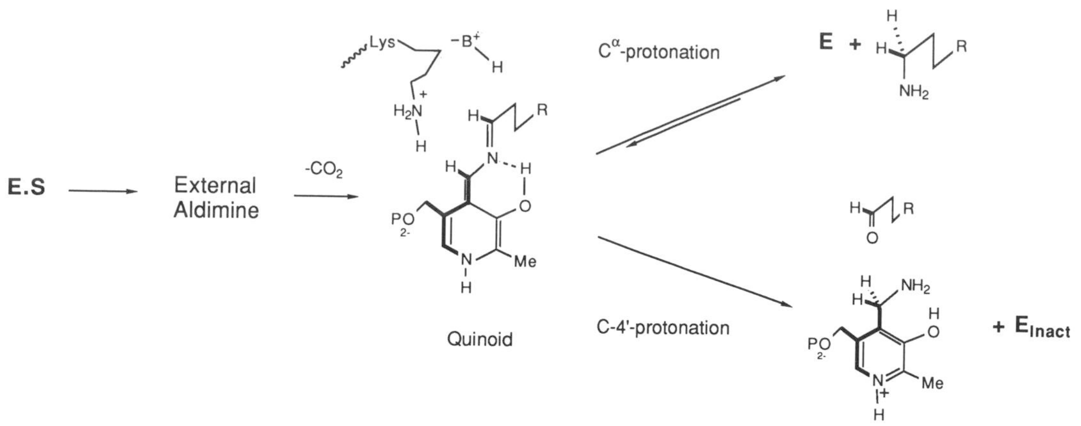

Scheme 3.

from the fern *Dryopteris filix-mas* and Streptomyces were identified as possessing suitable properties. Each of the proteins was purified to near homogeneity and was shown to be a homodimer ($M_r \approx 110000$). The two proteins showed similar activities at the optimum pH for V_{max} with methionine as substrate and were both able to catalyse the decarboxylation of a range of straight and branched chain hydrophobic amino acids (Stevenson *et al.* 1990 *a*, *b*).

To determine the stereochemical course of the decarboxylation reaction with respect to C^α, incubations of the enzyme with unlabelled and labelled substrate were conducted in deuterium oxide and protium oxide respectively, and the products were isolated and derivatized as their camphanamide derivatives. Comparison of the 360 MHz ^{1}H-NMR spectra of these samples with synthetic samples showed that each enzyme catalysed the decarboxylation of methionine with retention of configuration at C^α. Thus the enzymes displayed the more usual stereochemical course for α-amino acid decarboxylation (Stevenson *et al.* 1990 *a*, *b*).

When the stereochemical courses of the decarboxylation of several other substrates was determined for each methionine decarboxylase, retention of configuration was observed. Furthermore, each of the products showed a high chiral integrity. An interesting finding considering the diverse structures of the substrates.

Upon incubation of the enzymes with L-methionine in the absence of excess coenzyme, fern methionine decarboxylase, but not the *Streptomyces* enzyme, displayed activity loss. The addition of coenzyme to these incubations restored full activity and it appeared that enzyme was catalysing occasional abortive transamination events (see scheme 3). This explanation was verified in two different ways. First, [^{35}S]methionine was incubated with the enzyme and the expected transamination product, labelled 3-methylthiopropionaldehyde was detected and isolated. Second, PMP was isolated and characterized from similar incubations.

To determine the stereochemical course of the abortive transamination with the physiological substrate L-methionine, large-scale incubations were set up with the fern enzyme in which excess C-4' tritiated

coenzyme was added. The tritiated PMP generated during these incubations was isolated and purified and was treated with alkaline phosphatase to remove the phosphate ester group. The resulting tritiated pyridoxamine was incubated with freshly prepared apo-aspartate aminotransferase (which is known to exchange the 4'-*pro*-S hydrogen of pyridoxamine with the solvent, (Voet *et al.* 1973) and the tritium content of the water and the solid residue was determined at time intervals against controls containing racemic tritiated pyridoxamine. From the results it was evident that the tritium occupied the 4'-*pro*-R position of the pyridoxamine derived from the abortive transamination reaction and, therefore, that proton transfer to the C-4' position occurred from the *si*-face. Thus we had established, without ambiguity, that methionine decarboxylase showed the same stereochemical course for transamination of the coenzyme as AAT (Stevenson *et al.* 1990 *a*).

To check the generality of the stereospecificity for C-4' protonations in decarboxylases, the stereochemical course of the transamination reaction catalysed by *E. coli* glutamate decarboxylase (GAD) was determined with L-glutamic acid, the physiological substrate. The result was consistent with the emerging trend, that protonation occurred from the C-4' *si*-face of the coenzyme (Smith *et al.* 1991).

Given that a lysine residue in all PLP-dependent enzymes forms an aldimine linkage with the C-4' carbonyl group of the coenzyme, it seemed likely that this lysine residue was responsible for the protonation at C-4' and for the stereospecificity of the protonation. Thus one common feature for these different enzymes appeared to be emerging, that the ε-ammonium group of the active site was disposed on the 4'-*si*-face of the coenzyme. However, verification was needed.

(d) *Does lysine protonate at C^α?*

To determine whether the same or different conjugate acids protonate the quinoid intermediate at C^α and C-4' (see scheme 3), the enzyme was incubated with methionine and excess coenzyme in tritium oxide of high specific radioactivity. The tritium content of the 3-methylthio-1-aminopropane decarboxylation product, after crystallization to constant specific activity, was found to match that of the original incubation medium. The specific activity of the isolated PMP was, however, $\approx$ 40–50 times lower than that of the solvent. Taken at face value, the results showed that a monoprotic acid operates at C^α, while a polyprotic acid operates at C-4' (Akhtar *et al.* 1990). The latter result was rather pleasing, since it was consistent with the proposed role of the lysine ε-ammonium group. Furthermore, there could be no ambiguity in interpretation because the tritium content was so different to that of the solvent. The very slow rate of the abortive reaction and its predicted kinetic dominance by a chemical step, proton transfer, requires the observation of the intrinsic tritium isotope effect, statistically adjusted for the number of equivalent protons on the conjugate acid.

(e) Conjugate acid for C^α

The interpretation that a monoprotic acid operates at C^α, however, was not so clearcut since if the protonation step was followed by a very slow step, the proton transfer step would be brought into equilibrium, a situation in which the isolated product would display the specific activity of the solvent, regardless of the number of equivalent protons.

To determine whether the same or different bases or conjugate acids operated at C^α and C-4', a series of kinetic and kinetic isotope effect experiments were performed by using the normal decarboxylation, and the abortive transamination reactions, as probes (Akhtar *et al.* 1990). It was reasoned that if the partition ratio between decarboxylation and transamination events varied with pH, the result would be consistent with the operation of two bases whereas, a pH-insensitive partition ratio would be consistent with the operation of a single base or conjugate acid.

The experiments allowed a detailed analysis of the kinetic properties of the fern MD system and in particular, the kinetic resolution of pre-decarboxylation events (those leading-up to the first irreversible step, carbon dioxide desorption) and post-decarboxylation steps. First, the partition ratio (v_{am}/v_{ab}, where v_{am} and v_{ab} are the rates of amine and PMP formation, respectively) and V/K were pH-sensitive and showed sharp increases at pH 6.25. The profiles, therefore, showed that a group on the enzyme, in its un-protonated state, simultaneously suppresses the abort-ive reaction and enhances the physiological reaction. The results strongly suggested that the deprotonations of the ammonium groups of the lysine residue and the substrate (by the conjugate base of an acid possessing a molecular pK_a of $\approx$ 6.25) to generate the nucleophiles for each of the two transaldimination processes were responsible for the observed profiles (Akhtar *et al.* 1990).

Experiments conducted in deuterium oxide allowed the solvent isotope effect (SIE) for the partition ratio and for the abortive reaction to be determined. The SIE for the partition ratio increased sharply from 0.6 to 1.7 at pL 6.25 while that for the abortive reaction decreased from 1.7 to 0.7 indicating that a post-decarboxylation step on the normal reaction pathway

requires a catalytic group on the enzyme to be protonated. This step was probably quinoid protona-tion at C^α (Akhtar *et al.* 1990).

^{1}H-NMR (nuclear magnetic resonance) spectroscopic analysis of 3-methylthio-1-aminopropane isolated from incubations conducted in 50 (molar) % deuterium oxide at pL 4.8 and at pL 6.5, where the step for protonation of the quinoid at C^α was expected to be kinetically significant in the post-decarboxylation part of the reaction, showed that 50 % of deuterium was incorporated into the 1-*pro*-R position. The result shows that the proton donor was monoprotic and, in consideration of its kinetic titration, is the imidazolium sidechain of a histidine residue (Akhtar *et al.* 1990).

To determine if the two conjugate acids were disposed on the same faces of the coenzyme, or on different faces, the stereochemical courses for the decarboxylation were re-examined. The high chiral integrity of the monodeuteriated amine products for the diverse structural range of tested substrates suggested that the acids were disposed on the same face. If they were not, and the histidine was disposed on the C-4'-*re*-face, the bulkier substrates might have been expected to allow the lysine ε-ammonium group to occasionally protonate at C^α from the C-4'-*si*-face. Such events would give the deuteriated amine of the opposite configuration and would reduce the overall enantiomeric excess of the product. Given that the ability of the histidine to act as an acid would be diminished above pH 6.25 but, that of the ε-am-monium group would not, any protonation due to the action of lysine at C^α of the quinoid intermediate should be enhanced at pH 6.25 or higher. When the stereochemical courses for the decarboxylation of methionine and isoleucine were assessed at pL 7.0 in deuterium oxide the isolated (R)-monodeuteriated amines were completely pure. Thus as the lysine residue resides on the 4'-*si*-face of the coenzyme, the histidine must also.

4. COMPARISON OF ACTIVE SITE PEPTIDES

To compare the α-decarboxylases and SHMT to AAT, the X-ray crystal structures of the active-site of AAT were re-examined and the residues involved in binding to the coenzyme or to the substrate, or in catalysing specific chemical steps were identified (see table 2). Analysis of the regions of AAT which interact with the coenzyme showed that a tetrapeptide unit —S—X—X—K— contained two completely con-served residues, a serine (Ser 255) which forms a sidechain H-bond with the 5'-phosphate ester group of PLP, and the active site lysine (Lys 258), which forms the C-4' internal aldimine with the coenzyme and which also serves to shuttle protons between C^α and C-4' (table 1; see scheme 2).

An examination of all of the available active-site sequences for PLP-dependent enzymes was then undertaken and, in many, the residue equivalent to Ser 255 in AAT was a serine, asparagine, or a threonine residue, all H-bond donors. In particular, the active site tetrapeptides for decarboxylases and SHMT were well conserved (table 2) and contained the sequence —S(N,T)—X—H—K—, unless there were rationalizable mechanistic reasons for why they should not be conserved. For example, all decarboxylases known to catalyse the decarboxylation of (2S)-amino acids with retention of stereochemistry showed the conserved sequence but, *meso*-α,ω-diaminopimelate decarboxylase (which catalyses the decarboxylation at the (2R)-centre of its substrate with inversion of configuration) and mouse, rat, trypsanoma and yeast ornithine decarboxylases, some of which are the most unstable enzymes known, did not (Smith *et al.* 1991). Thus it appeared that for the decarboxylases which operate in a retentive mode, the serine, asparagine, or threonine residue might hydrogen bond to the 5′-phosphate ester group of the coenzyme, as in AAT, and that the histidine residue might be the catalytically important residue involved in quinoid protonation at C^α.

Given that the conformation of the homologous tetrapeptide section of each enzyme might be similar, it was instructive to note that histidine modifying reagents react with decarboxylases in the absence of substrate to give inactive proteins that possess unaltered binding affinities for their substrates. The result shows that the modified histidine residue is catalytically essential but, not important in the formation of the Michaelis complex (Dominici *et al.* 1985; Mishin & Sukhareva 1986). Gratefully, these interpretations are completely consistent with the role for histidine which we had proposed (Akhtar *et al.* 1990).

5. LOCATION OF DISTAL BINDING GROUPS

The final part of the comparison concerned the position of the distal substrate binding groups for decarboxylases and SHMT relative to those for the transaminases.

In decarboxylases the side chain could bind to a similarly positioned distal site as for transaminases (e.g. Arg 292, which is on the 5′ phosphate ester side on the coenzyme and binds the β- and γ-carboxylate groups of aspartate and glutamate in AAT) which would place the α-carboxyl group of the L-antipode of a substrate on the 4′-*re*-face of the coenzyme, or in an alternative mode which would place the carboxyl group of the 4′-*si*-face. Since typical decarboxylases catalyse the decarboxylation of L-amino acids with retention of configuration at C^α and as the histidine residue which protonates the quinoid intermediate at C^α appears to be disposed on the 4′-*si*-face (Akhtar *et al.* 1990), decarboxylases should possess a distal binding group that is located on the 3′-phenol side of the coenzyme ring. This is opposite to the situation for transaminases.

To check the validity of the analysis, some simple modelling experiments were performed. It was expected that minor changes to the distal binding groups in the external aldimine of AAT would give the correct disposition of catalytically functional groups and the correct external aldimine conformation for a decarboxylase.

Accordingly, the coordinates for the tetrapeptide —S—Y—A—K— and the substrate–coenzyme aldimine were excised from the X-ray crystal structure of

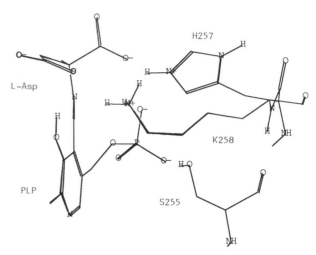

Figure 1. Result of the replacement of the methyl group of Ala 257 in the external aldimine form in chicken mitochondrial AAT by an imidazolium group, and rotation about the C^α-bond by 120°, and removal of the arginine residues. This conformation which was minimized, as described by Smith (1990), allows the formation of a hydrogen-bond between with the α-carboxylate group of the substrate and the N^3-proton of the nascent His 257 residue.

chicken mitochondrial AAT. Without changing any distances or torsional angles in the tetrapeptide, the C^α-bond of the substrate–aldimine was rotated at 120° so that the α-carboxyl group of the substrate was on the 4′-*si*-face of the coenzyme at the angle required for maximum stereoelectronic assistance (Dunathan 1966). Without any further change, the position of Ala 257 was examined. The sidechain pointed directly towards the substrate. The alanine residue, in keeping with the decarboxylase motif, was then altered by modelling to histidine and the interactions were energy minimized by using Macromodel (see figure 1). The α-carboxylate group formed an H-bond with the protonated histidine ε-N-atom and the entire system appeared to be optimally set up to catalyse decarboxylation. (These interactions were also modelled with all of the amino acid residues which interact with the external aldimine of AAT in place except Arg 386. Here the configuration of the substrate was inverted to the D-antipode so that its α-carboxylate group was disposed on the 4′-*si*-face of the coenzyme, see Smith (1990)).

From this analysis it is expected that the histidine residue does not serve as a catalytic base in the decarboxylation step but, in its protonated form, ensures that the conformation of the C^α—N bond is close to 90° to the plane of the conjugated π-electron system. Following decarboxylation, and generation of the quinoid intermediate, the imidazolium sidechain then protonates the quinoid from C-4′-*si*-face at C^α to give the product aldimine. The findings reviewed and presented here allow a very detailed comparison of PLP-dependent enzymes and the rationalization of the stereochemical courses of most of the reactions which have been studied to date but, not those for SHMT, *vide infra*. The major difference between transaminases and decarboxylases appears to be the conformation of the substrate aldimine C^α—N bond which is controlled

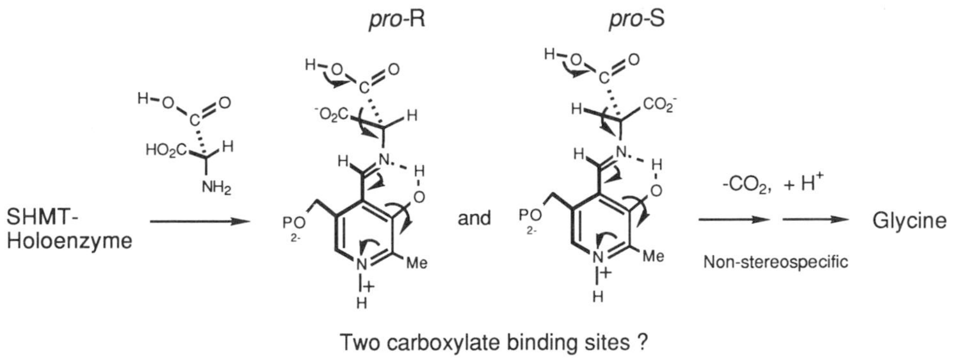

Two carboxylate binding sites ?

Scheme 4.

by sidechain binding and, in decarboxylases, by the presence of the positively charged imidazolium side-chain of a histidine residue which interacts electro-statistically with the α-carboxylate group. Our cumulative results show that histidine serves two other functions; as a base in the deprotonation of the ammonium group of the substrate and lysine residue immediately before transaldimination reactions, and; as a proton donor for the quinoid intermediate at C^α. These findings are completely in accord with the Dunathan postulate and show that it should be possible to design new decarboxylases starting from AAT (Smith 1990). Researches directed towards these goals are underway.

6. SERINE HYDROXYMETHYLTRANSFERASE

SHMT shows many similarities to α-amino acid decarboxylases, including the ability to catalyse decar-boxylation and, the S(T,N)—X—H—K— tetrapep-tide decarboxylase motif. However, the enzyme is unusual in that it shows a low regard for reaction-type specificity with α-amino acid substrates and catalyses many of these reactions non-stereospecifically (Thomas *et al*. 1990).

As noted earlier, Palekar *et al*. (1973) showed that in tritiated water the decarboxylation of aminomalonic acid by SHMT gave both (2R)- and (2S)-tritiated glycine and that incubation with specific carboxyl-labelled [14C]aminomalonate confirmed that the re-action was non-stereospecific. To explain the apparent lack of stereospecificity, Palekar *et al*. (1973) proposed that the substrate might bind in two conformations at the active site of the enzyme, such that each of the two carboxyl groups was positioned correctly for de-carboxylation (scheme 4). If each conformer was equally populated and, the decarboxylation and subsequent protonation steps occurred stereospecific-ally for each form, then apparent non-stereospecific decarboxylation would be observed. However, if the enzyme was able to catalyse the racemization of the substrate before decarboxylation, the same observa-tions might have been expected. The latter scenario, before racemization, may have seemed unlikely to Palekar *et al*. in view of their earlier finding that PLP-dependent aspartate β-decarboxylase catalysed the stereospecific decarboxylation of aminomalonate

(Palekar *et al*. 1971). To unravel the mechanistic and stereochemical ambiguities, a new substrate was sought, preferably one that could not racemize (Thomas *et al*. 1990).

2-Amino-2-methylmalonic acid was prepared and was tested as a substrate for cytosolic rabbit liver SHMT. At pH 7.5 significant enzymic decarboxylation occurred relative to control incubations containing PLP but, no enzyme. The formation of alanine was detected by thin layer chromatography, and the product was isolated, and characterized by H-NMR spectroscopy (Thomas *et al*. 1990).

The absolute stereochemistry of the alanine was determined by incubating aliquots of the reaction solution at various time intervals with enzyme cocktails containing either D-amino acid oxidase and lactate dehydrogenase or L-alanine dehydrogenase. Analysis revealed that only (2R)-alanine was formed initially (note: SHMT is able to catalyse the racemization of (2R)-alanine upon prolonged incubation (Shostack & Schirch 1988)). Hence, one stereochemical aspect of the decarboxylation had been solved, and whichever carboxyl group was lost, the resulting quinoid in-termediate was protonated from the *si*-face at C^α to give (2R)-alanine.

To facilitate experiments to determine whether the cleavage of a unique carboxyl group or the cleavage of both carboxyl groups could give rise to a quinoid intermediate a synthesis of the enantiomers of [1-13C]2-amino-2-methylmalonic acid was devised by using Schollkopf *bis*-lactim ether methodology (Thomas *et al*. 1990; Thomas & Gani 1991).

The labelled chiral aminomalonates were each incubated with SHMT and PLP at pH 7.5 and the resulting alanines were isolated. Examination of the products by ¹H- and ¹³C-NMR spectroscopy showed that the (2R)-2-amino-2-methylmalonate gave un-labelled alanine (δ_H 1.21 p.p.m.; d, $J_{H-2,3} = 6.8$ Hz, in D₂O at pH 10) whereas the (2S)-enantiomer gave [1-¹³C]alanine (δ_H 1.21 p.p.m.; dd, $J_{H-2,3} = 6.8$ Hz, $J_{H-2,C-1} = 4$ Hz). Thus the *pro*-R carboxyl group of the substrate was lost during the decarboxylation. To-gether with the finding that D-alanine was the decarboxylation product, it was evident that replace-ment of the *pro*-R carboxyl group by a proton occurred with retention of configuration (Thomas *et al*. 1990). The same stereochemical course was observed for the *E. coli* enzyme (Thomas 1990).

Phil. Trans. R. Soc. Lond. B (1991)

Interestingly, the *pro*-R carboxyl group of 2-amino-2-methylmalonic acid is expected to occupy the same position at the active-site of the enzyme as the hydroxymethyl group of the physiological substrate, L-serine, in complete accord with the Dunathan postulate and our model for the active-site of PLP-dependent α-decarboxylase enzymes.

In conclusion, our work with SHMT showed that the findings of Palekar *et al.* (1973) are best rationalized in terms of the prior enzymic or non-enzymic racemization of the 2-aminomalonic acid. Indeed, in our hands, glycine isolated from the incubation of cytosolic SHMT with 2-aminomalonic acid at pL 6.0 in deuterium oxide contained more than 1.8 equivalents of deuterium at the C-2 position. Examination of the [^{1}H] and [^{2}H]NMR spectra of the camphanamide derivatives revealed that the deuterium was almost evenly distributed between the 2-*pro*-R and 2-*pro*-S positions of the glycine, implying that racemization occurs before decarboxylation. In accord with this conclusion, control experiments, which were done under identical conditions, but which contained no enzyme showed that the exchange of protium from the substrate was very rapid with half lives for exchange of $\approx$ 10–20 min over the pH range of interest (Thomas 1990).

When incubations containing high concentrations of SHMT and low concentrations of 2-aminomalonic acid were performed in deuterium oxide, a highly stereoselective decarboxylation reaction was observed (Thomas 1990). The predominant product was (2S)[2-^{2}H]glycine and, again, the major side product was the dideuterio isotopomer. Thus there is no evidence to suggest that SHMT catalyses non-stereospecific decarboxylation for any substrate (see scheme 4). The implication of these findings in the light of the model for the α-decarboxylases is that the conserved histidine residue in SHMT (see table 2) should serve as an H-bond donor for the β-hydroxyl group in the retro-aldol cleavage of L-serine. The high level of activity observed for the cleavage of serine catalysed by an *E. coli* His-Asn mutant SHMT (Shostack & Schirch 1988) is in accord with the notion that the side chain of the histidine should be protonated and should not serve as a base. Furthermore, it appears that the binding site for the α-carboxyl group of L-serine, and D-alanine (an abortive transamination substrate) and for the *pro*-S carboxyl groups of 2-aminomalonic acid decarboxylation substrates should be disposed on the 5'-phosphate ester side of the coenzyme, in a similar but slightly nearer position to that of Arg 292 in AAT. Reactions of L-alanine with SHMT cannot be rationalized at the present time and extensive further studies of SHMT will be required to elucidate the kinetic influence and mechanistic role of the formaldehyde acceptor, tetrahydrofolic acid, in the physiological reaction.

I thank my co-workers, the researchers who actually did the work, Dr M. Akhtar, Dr D. E. Stevenson, Dr D. M. Smith and Dr N. R. Thomas, and for their constant enthusiasm and dedication. We are all indebted to Professor J. Jansonius for providing the X-ray coordinates for chicken mitochondrial AAT, to Professor V. Schirch for providing SHMT enzymes and to S. Chamberlin for help with running Macromodel.

We thank the SERC for financial support and for studentships to D.M.S. and N.R.T. and the Royal Society for a Royal Society University Fellowship to D.G. for 1983–88.

REFERENCES

Akhtar, M., Stevenson, D. E. & Gani, D. 1990 Fern L-methionine decarboxylase: kinetics and mechanism of decarboxylation and abortive transamination. *Biochemistry* **29**, 7648–7660.

Dominici, P., Tancini, B. & Voltattorni, C. B. 1985 Chemical modification of pig kidney 3,4-dihydroxyphenylalanine decarboxylase with diethyl pyrocarbonate: evidence for an essential histidine residue. *J. Biol. Chem.* **260**, 10583–10589.

Dunathan, H. C. 1966 Conformation and reaction specificity in pyridoxal phosphate enzymes. *Proc. natn. Acad. Sci. U.S.A.* **55**, 712–716.

Gani, D. 1985 Enzyme chemistry. *Ann. Rep. Prog. Chem. B* **82**, 287–310.

Gani, D. 1990 Pyridoxal-dependent systems. In *Comprehensive medicinal chemistry*, vol. 2. (ed. P. G. Sammes), pp. 213–254. Oxford: Pergamon Press.

Kelland, J. G., Palcic, M. M., Pickard, M. A. & Vederas, J. C. 1985 Stereochemistry of lysine formation by meso-diaminopimelate decarboxylase from wheat germ: use of proton–^{13}C NMR shift correlation to detect stereospecific labelling. *Biochemistry* **24**, 3263–3267.

Kirsch, J. F., Eichele, G., Ford, G. C., Vincent, M. G., Jansonius, J. N., Gehring, H. & Christen, P. 1984 Mechanism of action of aspartate aminotransferase proposed on the basis of its spatial structure. *J. Molec. Biol.* **174**, 497–525.

Mishin, A. A. & Sukhareva, B. S. 1986 Glutamate decarboxylase from E. coli: catalytic role of a histidine residue. *Dokl. Acad. Sci. S.S.S.R.* **290**, 1268–1271.

Palekar, A. G., Tate, S. S. & Meister, A. 1971 Stereospecific decarboxylation of specifically labelled ^{13}C-carboxyl aminomalonic acids by L-aspartate β-decarboxylase. *Biochemistry* **10**, 2180–2182.

Palekar, A. G., Tate, S. S. & Meister, A. 1973 Rat liver aminomalonate decarboxylase: identity with cytoplasmic serine hydroxymethyltransferase and allothreonine aldolase. *J. biol. Chem.* **248**, 1158–1167.

Shostack, K. & Schirch, V. 1988 Serine hydroxymethyltransferase: mechanism of the racemization and transamination of D & L-alanine. *Biochemistry* **27**, 8007–8014.

Smith, D. M. 1990 Structural and mechanistic studies of pyridoxal 5'-phosphate dependent enzymes. Ph.D. thesis, University of Southampton.

Smith, D. M., Thomas, N. R. & Gani, D. 1991 A comparison of pyridoxal 5'-phosphate dependent decarboxylase and transaminase enzymes at a molecular level. *Experentia.* (In the press.)

Stevenson, D. E., Akhtar, M. & Gani, D. 1990 *a* L-methionine decarboxylase from *Dryopteris filix-mas*: purification, characterization, substrate specificity; abortive transamination of the coenzyme and the stereochemical courses of substrate decarboxylation and coenzyme transamination. *Biochemistry* **29**, 7631–7647.

Stevenson, D. E., Akhtar, M. & Gani, D. 1990 *b Streptomyces* L-methionine decarboxylase: purification and properties of the enzyme; stereochemical course of substrate decarboxylation. *Biochemistry* **29**, 7660–7666.

Sukhareva, B. S. & Braunstein, A. E. 1971 Nature of the interaction of *Escherichia coli* glutamate decarboxylase with substrate and substrate analogues. *J. molec. Biol. S.S.R.* **51**, 302–317.

Thomas, N. R. 1990 Synthesis of isotopically labelled

substrates and their use in stereochemical and mechanistic studies of enzyme action. Ph.D thesis, University of Southampton.

Thomas, N. R. & Gani, D. 1991 Synthesis of (2R)- and (2S)-[1-^{13}C]-2-amino-2-methylmalonic acid: chiral substrates for serine hydroxymethyltransferase. *Tetrahedron.* **47**, 497–506.

Thomas, N. R., Schirch, V. & Gani, D. 1990 Synthesis of (2R)- and (2S)-[1-^{13}C]-2-amino-2-methylmalonic acid, probes for the serine hydroxymethyltransferase reaction: stereospecific decarboxylation of the 2-*pro*-R carboxyl group with the retention of configuration. *J. chem. Soc. chem. Comm.* 400–402.

Voet, J. G., Hindenlang, D. M. T., Blanck, J. J., Ulevitch, R. J., Kallen, R. G. & Dunathan, H. C. 1973 Stereochemistry of pyridoxal phosphate enzymes: absolute stereochemistry of cofactor C$_4'$ protonation in the transamination of holoserine hydroxymethylase by D-alanine. *J. biol. Chem.* **248**, 841–842.

Yamada, H. & O'Leary, M. H. 1977 A solvent isotope effect probe for enzyme mediated proton transfers. *J. Am. chem. Soc.* **99**, 1660–1661.

Mechanisms of sulphate activation and transfer

GORDON LOWE

The Dyson Perrins Laboratory, Oxford University and the Oxford Centre for Molecular Sciences, South Parks Road, Oxford OX1 3QY, U.K.

SUMMARY

Sulphation of natural products is a widespread phenomenon. Inorganic sulphate is transported into cells and activated by ATP sulphurylase, an enzyme that has been studied by kinetic and stereochemical methods. It has been shown that the enzyme catalyses the displacement of inorganic pyrophosphate by inorganic sulphate from P_α of ATP by a direct 'in line' mechanism. The adenosine 5'-phosphosulphate formed is then phosphorylated at the 3' position by APS kinase to give 3'-phosphoadenosine 5'-phosphosulphate, the common sulphating species in biology.

A general strategy for the synthesis of chiral [$^{16}O^{17}O^{18}O$]-sulphate esters has been established and a method developed for their stereochemical analysis by using Fourier Transform Infrared Spectroscopy. The stereochemical course of an aryl sulphotransferase from *Aspergillus oryzae* has been shown to proceed with retention of configuration at sulphur, supporting a ping pong type mechanism with a sulpho-enzyme intermediate on the reaction pathway.

INTRODUCTION

Sulphate half esters are found among all the major classes of natural products, i.e. nucleotides, peptides and proteins, polysaccharides, steroids and lipids. Since the classical study of Baumann (1876) well over a century ago, in which phenol administered to a patient was excreted as phenyl sulphate it has been clear that sulphation is used as a detoxification mechanism. Sulphation, however, plays other important functional roles among natural products. A number of secreted peptides and proteins (e.g. fibrinogen, fibrinopeptide A and B, cholecystokinin, hirudin, etc.) possess sulphated tyrosine residues. Thus fibrinogen, a soluble plasma protein, on proteolysis with thrombin releases fibrinopeptide A (18 residues) and B (20 residues), and the fibrin monomers formed spontaneously associate into insoluble fibres. Fibrinopeptides A and B both contain a sulphated tyrosine residue as well as aspartate and glutamate residues. The presence of these regions of high negative charge in fibrinogen prevents aggregation. Interestingly, the anticoagulant hirudin produced by the medicinal leech (*Hirudo medicinalis*), is a 65-residue peptide containing a negatively charged C-terminus, which includes a sulphated tyrosine residue at position 63. Hirudin is a potent inhibitor of thrombin and so prevents blood clotting at the site of incision thus allowing blood to be drawn freely. A recent report on the X-ray structural analysis of the complex of recombinant hirudin and human α-thrombin shows that the negatively charged tail binds at a highly positively charged region of thrombin remote from the active site (Rydel *et al.* 1990). The sulphated polysaccharide heparin also binds to this site in α-thrombin. The sulphate group found in proteoglycans (e.g. chondroitin sulphate, dermatan sulphate, etc.) ensures that the polysaccharide chains are separated and in extended conformations because of electrostatic repulsion. The function of these molecules appears to be to create a gel around cells with the ability to disperse shock. They are found in cartilage, arterial walls and connective tissue (Schubert & Hamerman 1968). Sulphate esters of steroids are important intermediates in the biosynthesis of steroidal hormones (e.g. progesterone, estrone, androstenolone, cortisone, etc.) (Bernstein & Solomon 1970). Steroid sulphatase deficiency leads to excretion of large amounts of steroidal sulphates leading to problems in pregnancy and childbirth (Shapiro *et al.* 1977). Sulpholipids are found in place of phospholipids in the membranes of some bacteria, clearly showing that sulphate monoesters can play a similar role to phosphate monoesters.

A particularly interesting function of sulphate esters is found in the mustard oil glycosides (e.g. sinigrin) present in *Brassica* species. When the thioglycoside bond is cleaved by myrosinase the sulphated thiohydroxamate undergoes a spontaneous Lossen-type rearrangement to give a volatile isothiocyanate, which is responsible for the characteristic biting taste and pungent odour of ground mustard seeds (Kjaer 1960, 1961).

R = allyl for sinigrin.

The pre-emergent herbicide 2-(2,4-dichlorophenoxy)ethyl sulphate (Crag herbicide, also known as Herbon) is inactive until the sulphate ester is hydrolysed by sulphatases produced by soil microorganisms,

Phil. Trans. R. Soc. Lond. B (1991) **332**, 141–148

Printed in Great Britain

[35]

141

in particular *Pseudomonas* species. The sulphate group makes the pro-herbicide water soluble and therefore easy to apply as well as ensuring a slow release of the active agent (Lillis *et al.* 1983).

ACTIVATION OF INORGANIC SULPHATE

Inorganic sulphate is used in the biosynthesis of sulphate half esters, but this inert, water soluble divalent ion has to be transported into cells and then enzymically activated. The structure of the sulphate binding protein that actively transports sulphate ions into *Salmonella typhimurium* has recently been determined at 2 Å† resolution (Pflugrath & Quiocho 1988). Although inorganic sulphate is an exceedingly poor nucleophile in aqueous solution, it behaves as such in the activation process. Robbins & Lipmann (1956, 1957, 1958) first isolated biologically activated sulphate and correctly characterized it as the nucleotide 3′-phosphoadenosine 5′-phosphosulphate (PAPS). It was subsequently synthesized by Baddiley *et al.* (1957).

3′-Phosphoadenosine 5′-phosphosulphate is synthesized enzymically from adenosine 5′-triphosphate and inorganic sulphate. ATP sulphurylase catalyses the first step in the synthesis in which adenosine 5′-phosphosulphate (APS) is formed.

$$\text{ATP}^{4-} + \text{SO}_4^{2-} + \text{H}^+ \rightleftharpoons \text{APS}^{2-} + \text{PP}_i^{3-} \rightarrow 2\text{P}_i.$$

The equilibrium constant for the formation of APS however is very unfavourable, $\Delta G_0'$ being about $+9.5$ kcal mol^{-1} at 5 mM Mg^{2+}, 25 °C and pH 7 (Robbins & Lipmann 1958; Akagi & Campbell 1962). As $\Delta G_0'$ for the hydrolysis of ATP to AMP and PP$_i$ under the same conditions is -10 kcal mol^{-1} (Gwynn *et al.* 1974), $\Delta G_0'$ for hydrolysis of the phosphosulphate anhydride bond must be approximately -19.5 kcal mol^{-1}. Because of the unfavourable $\Delta G_0'$ for the formation of APS, if no other reactions were involved the concentration of the activated sulphate would be very low. The liberated inorganic pyrophosphate, however, is hydrolysed by inorganic pyrophosphatase, and the free energy locked in this phosphoanhydride bond released. $\Delta G_0'$ for this reaction is about -4.5 kcal mol^{-1} in the presence of excess Mg^{2+} at pH 7.

The third enzyme involved in the biosynthesis of PAPS is APS kinase which uses another molecule of ATP to phosphorylate the 3′-hydroxy group of APS. $\Delta G_0'$ for breaking the phosphoanhydride bond is between -4 and -5 kcal mol^{-1} bringing the overall equilibrium constant for the formation of PAPS from inorganic sulphate close to unity.

$$\text{APS}^{2-} + \text{ATP}^{4-} \rightleftharpoons \text{PAPS}^{4-} + \text{ADP}^{3-} + \text{H}^+.$$

In bacteria and fungi the activated sulphate group in PAPS can be reduced by NADPH first to inorganic sulphite with the enzyme PAPS reductase and then to hydrogen sulphide with sulphite reductase which is used for the biosynthesis of divalent sulphur containing compounds. PAPS is the common sulphating agent in biology.

† 1 Å = 10^{-10} m = 10^{-1} nm.

ATP SULPHURYLASE

ATP sulphurylase has been extensively studied. Initial velocity, product inhibition, dead-end inhibition, and isotope exchange studies have shown that ATP sulphurylase follows an ordered sequential kinetic mechanism, MgATP adding before sulphate and MgPP$_i$ dissociating before APS (Wilson & Bandurski 1958; Stadtman 1973; Tweedie & Segel 1971; Shoyab *et al.* 1972; Farley *et al.* 1976, 1978). MgATP analogues (ATP, CaATP, CrATP, or Mg αβ-methylene ATP), however, were shown to be incapable of inducing the binding of sulphate, and so it was concluded that either the structural requirements for the metal ion-nucleotide complex are so stringent that only MgATP can induce the conformational change that forms the sulphate binding site, or the mere binding of a metal ion-nucleotide complex is insufficient to promote the binding of sulphate. The latter explanation was favoured, with the implication that the MgATP must be catalytically cleaved to E-AMP. MgPP$_i$ before the sulphate binding site formed, and that the adenylated enzyme then bound inorganic sulphate and reacted with it. This conclusion, however was shown by stereochemical methods to be untenable as displacement at P_α of ATP by sulphate ion catalysed by ATP sulphurylase proceeds with inversion of configuration at phosphorus (Bicknell *et al.* 1982). Since our work was reported, Seubert *et al.* (1983, 1985) have shown that if ATP sulphurylase is rigorously purified the earlier observations, which had led to the proposal of an adenylated enzyme intermediate, could not be repeated. The kinetic and stereochemical evidence now clearly indicate that the activation of inorganic sulphate catalysed by ATP sulphurylase occurs by a direct 'in line' displacement at P_α of ATP.

SYNTHESIS AND STEREOCHEMICAL ANALYSIS OF CHIRAL [^{16}O,^{17}O,^{18}O] SULPHATE ESTERS

To study the stereochemical course of sulphuryl transfer reactions a general strategy was needed for the synthesis of chiral [^{16}O,^{17}O,^{18}O]sulphate monoesters of known absolute configuration as well as a method for their stereochemical analysis.

The general strategy for synthesis was to take a chiral 1,2- or 1,3-diol, convert it into the diastereoisomeric cyclic sulphites with [^{18}O]thionyl chloride and then oxidise the sulphite esters to the cyclic sulphate diesters with ruthenium [^{17}O]tetroxide. Regiospecific ring opening of the cyclic [^{17}O,^{18}O]sulphate diesters should give chiral [^{16}O,^{17}O,^{18}O]sulphate monoesters of known absolute configuration (Lowe & Salamone 1984). Nuclear magnetic resonance (N.M.R.) spectroscopy, which had been successfully used for the analysis of chiral [^{16}O^{17}O^{18}O]phosphate monoesters (Lowe 1983), did not commend itself for the analysis of chiral [^{16}O,^{17}O,^{18}O]sulphate monoesters as none of the isotopes of sulphur possesses a nuclear spin quantum number of $\frac{1}{2}$. Moreover, the exocyclic oxygen atoms in a cyclic sulphate diester cannot be alkylated to render

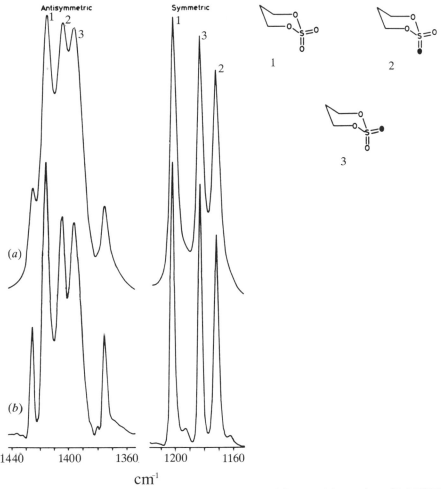

Figure 1. (*a*) The FTIR spectrum of the antisymmetric and symmetric $> SO_2$ stretching region of 2,2-[^{18}O]dioxo-1,3,2-dioxathiane and (*b*) the same spectrum after deconvolution with an enhancement factor of 1.68 and line-width at half height of 5 cm^{-1} for the antisymmetric $> SO_2$ stretching region and an enhancement factor of 1.50 and line-width at half height of 4 cm^{-1} for the symmetric $> SO_2$ stretching region. The spectra were measured on a Perkin-Elmer 1750 FTIR spectrometer at a resolution of 1 cm^{-1}.

Table 1. *The effect of oxygen isotopic substitution on the symmetric and antisymmetric* $> SO_2$ *stretching frequencies* (cm^{-1}) *of* (*4R*)*-4-methyl-2,2-dioxo-1,3,2-dioxathianes*

(Δ is the isotope shift from the $> SO_2$ frequency.)

isotope		symmetric stretching frequency	Δ for symmetric stretching mode	antisymmetric stretching frequency	Δ for antisymmetric stretching mode
axial	equatorial				
^{16}O	^{16}O	1201	—	1414	—
^{16}O	^{17}O	1192	9	1401	13
^{17}O	^{16}O	1186	15	1407	7
^{16}O	^{18}O	1183	18	1392	22
^{18}O	^{16}O	1172	29	1401	13
^{17}O	^{18}O	1170	31	1384	30
^{18}O	^{17}O	1163	38	1389	25
^{18}O	^{18}O	1157	44	1378	36

them chemically distinguishable as was possible with cyclic phosphate diesters. The analysis of chiral [$^{16}O,^{17}O,^{18}O$]sulphate monoesters presented, therefore, a new conceptual problem.

The frequency of an infrared (IR) vibrational mode is markedly affected if isotopic substitution occurs in the functional group responsible. However, the magnitude of the isotope shift is difficult to predict except

in the simplest of molecules, but the shift caused by replacing ^{16}O with ^{18}O in a functional group could be up to 40 cm^{-1} (Pinchas & Laulicht 1971). We expected that the generalized anomeric effect (Kirby 1983) of the axial lone pairs of the ring oxygens on the axial exocyclic oxygen in 2,2-dioxo-1,3,2-dioxathianes, would ensure that the frequencies of the symmetric and antisymmetric $> SO_2$ vibrational modes were depen-

Phil. Trans. R. Soc. Lond. B (1991)

dent on the location of the heavy oxygen isotope. If so, the isotope effect could form the basis of a method for the analysis of chiral [^{16}O,^{17}O,^{18}O]sulphate monoesters.

To explore the influence of oxygen isotopic substitution at the axial and equatorial sites of six-membered cyclic sulphate esters, 2,2-[^{18}O]dioxo-1,3,2-dioxathiane was prepared from 2-oxo-1,3,2-dioxathiane by oxidation with ruthenium [^{18}O$_4$]tetraoxide (Lowe & Salamone 1983), since this should exist as an equimolar mixture of chair conformations (neglecting the equilibrium isotope effect) with ^{18}O in the axial and equatorial sites.

The FTIR spectrum is shown in figure 1a (the ^{18}O site is only about 60% enriched). It is evident from this spectrum that the ^{18}O shifts in both the symmetric and antisymmetric stretching modes of the $> SO_2$ group are conformationally dependent. It is also apparent that the isotope shift is greater and the intrinsic linewidth smaller in the symmetric stretching mode; consequently this vibrational mode should be the most useful for analytical purposes. Separate deconvolution of these two spectral regions gave the resolution-enhanced spectrum shown in figure 1b.

To assign the conformations responsible for the two symmetric and two antisymmetric $> S[^{16}O,^{18}O]$ absorption bands in 2,2-[^{18}O]dioxo-1,3,2-dioxathiane, and to confirm this observation, the *cis-* and *trans-*cyclic sulphite esters obtained by treating (3R)-butane-1,3-diol with thionyl chloride were oxidized with ruthenium [^{18}O$_4$]tetraoxide to give the isotopomeric cyclic [^{18}O]sulphate esters. Since this oxidation is known to proceed with retention of configuration at sulphur (Lowe & Salamone 1983), the *cis-*sulphite must give the cyclic (R_S)-[^{18}O]sulphate 1 and the *trans-*sulphite must give the cyclic (S_S)-[^{18}O]sulphate 2. As expected, the (R_S)-isotopomer 1 possesses only one symmetric (1172 cm^{-1}) and one antisymmetric (1401 cm^{-1}) $> S[^{16}O,^{18}O]$ stretching vibration and likewise the (S_S)-isotopomer 2 possesses only one symmetric (1183 cm^{-1}) and one antisymmetric (1392 cm^{-1}) $> S[^{16}O,^{18}O]$ stretching vibration, since the conformation with the methyl group equatorial should be preferred. Although these values differ slightly (1 cm^{-1}) from those observed for 2,2-[^{18}O]dioxo-1,3,2-dioxathiane, the assignments shown in figure 1 are unambiguous.

The three stable oxygen isotopes, ^{16}O, ^{17}O, and ^{18}O can be arranged to give nine exocyclic isotopomers of (4R)-4-methyl-2,2-dioxo-1,3,2-dioxathiane. The

^{17}O^{17}O isotopomer will not appear in the FTIR analysis but the other eight isotopomers will and have been prepared and their FTIR spectra determined. The frequencies of the symmetric and antisymmetric stretching modes for each isotopomer are shown in table 1. As each of the diastereoisotopomeric pairs are distinguishable, IR spectroscopy should provide a means for analysing chiral [^{16}O,^{17}O,^{18}O]sulphate monoesters after stereospecific cyclization to a chirally substituted six-membered cyclic sulphate.

(S_S)- and (R_S)-(1R)-3-Hydroxy-1-methylpropyl [^{16}O,^{17}O,^{18}O]sulphates 7 and 8 were prepared as outlined in figure 2a. [^{18}O]Thionyl chloride, prepared from sulphur [^{18}O$_2$]dioxide (99 atom% ^{18}O) and phosphorus pentachloride, was used to prepare the *cis-* and *trans-*(4R)-4-methyl-2−[^{18}O]oxo-1,3,2-dioxathianes 3 and 4 from (3R)-butane-1,3-diol. The separated diastereoisomers were oxidized with ruthenium [^{17}O$_4$]-tetraoxide (prepared *in situ* from ruthenium dioxide, sodium periodate, and [^{17}O]water). Because this oxidation is known to proceed with retention of configuration at sulphur (Lowe & Salamone 1983), the *cis-*[^{18}O]sulphite 3 gives the (2S)-compound 5 and the *trans-*[^{18}O]sulphite 4 gives the (2R)-compound 6.

The hydrolytic cleavage of 4-methyl-2,2-dioxo-1,3,2-dioxathiane has been extensively studied, but no conditions had been reported which gave exclusive cleavage of the primary C−O bond (Lichtenberger 1948; Lichtenberger & Durr 1956). Ammonia in methanol, however, gave the desired mode of ring cleavage, the primary amines being isolated virtually quantitatively. The corresponding primary alcohols 7 and 8 were obtained by treatment with nitrous acid in 83% yield.

It was now necessary to develop a stereospecific method for the cyclization of the enantiomeric [^{16}O,^{17}O,^{18}O]sulphate monoester 7 and 8. Lack of precedent for the formation of cyclic sulphate esters from acyclic sulphate monoesters led to the exploration of several possible reagents. Only two were found, namely trifluoromethanesulphonic anhydride and sulphuryl chloride, the latter giving slightly better yields.

To investigate whether there was any isotope exchange during cyclization, (1R)-3-hydroxy-1-methylpropyl[^{18}O]sulphate was prepared and cyclized with sulphuryl chloride. The chemical ionization mass spectrum (NH$_3$) of the cyclic sulphate obtained revealed a molecular ion at m/z 172 only (relative molecular mass, M_r for C$_4$H$_8$SO$_3^{18}$O·NH$_4^+$ is 172), suggesting that cyclization had occurred by activating the primary alcohol followed by intramolecular displacement by the sulphate monoester (figure 2b). This mode of cyclization was confirmed by the natural abundance ^{13}C NMR spectrum of the cyclic sulphate which showed C-1 to be split into two resonances at δ 71.784 and 71.749 p.p.m., the endocyclic ^{18}O causing an upfield shift of 0.035 p.p.m. as expected (Risley & Van Etten 1979; Vederas 1980; Hansen 1983) and in a 2:1 ratio of intensity after correcting for the ^{18}O enrichment of the sulphate monoester; thus no loss of isotope had occurred. It was now of interest to investigate the FTIR spectrum of the mixture of isotopomeric cyclic sulphate esters. As expected, three absorption bands

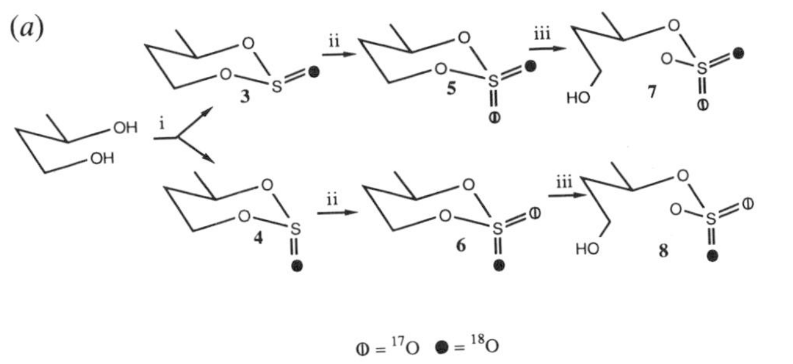

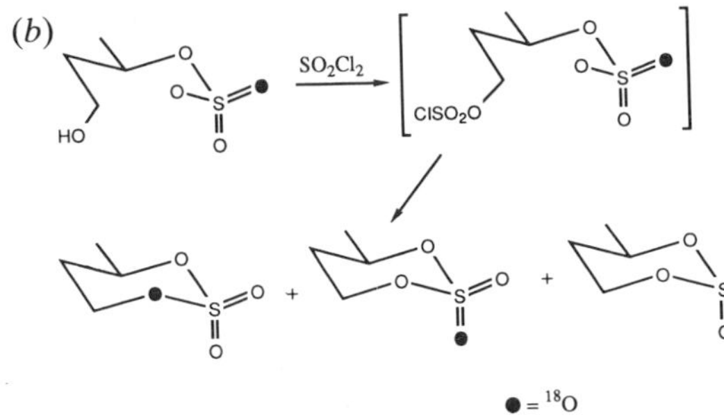

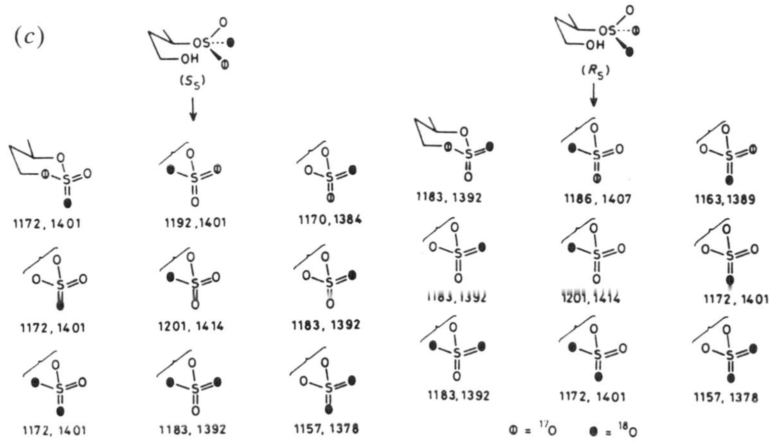

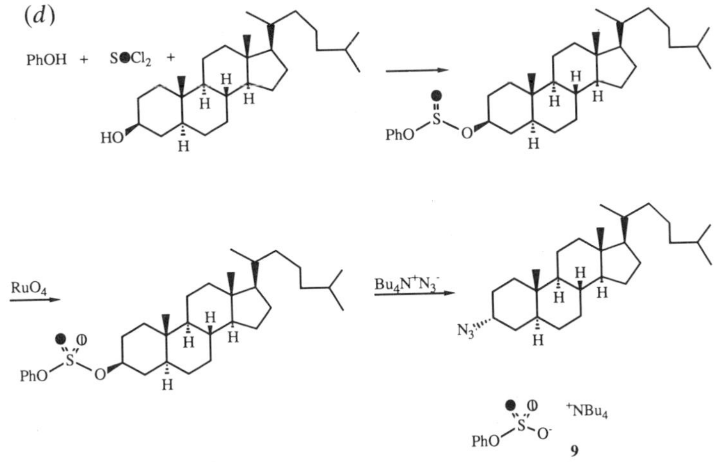

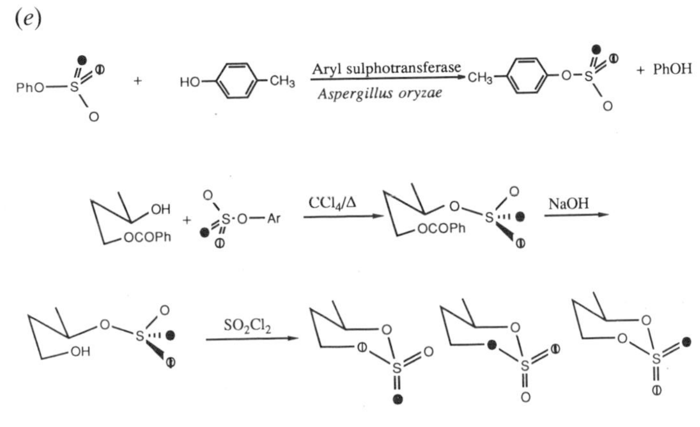

Figure 2 *a–e*. (*a*) The synthesis of the (S_S)- and (R_S)-sulphates **7** and **8**. Reagents: i, $S^{18}OCl_2$, C_5H_5N; ii, $Ru^{17}O_4$ (from RuO_2, $NaIO_4$, and $H_2{}^{17}O$); iii, NH_3, MeOH; $NaNO_2$, aq. AcOH. (*b*) The mechanism of cyclization of (1*R*)-3-hydroxy-methylpropyl sulphate. (*c*) The cyclization of the (S_S)- and (R_S)-sulphates **7** and **8** with retention of configuration at sulphur. If the three isotopes were fully enriched only the first three isotopomers of each set would be formed, but in practice the '^{17}O site' contains substantial amounts of ^{16}O and ^{18}O and consequently nine isotopomers should be formed for each chiral [$^{16}O,^{17}O,^{18}O$]sulphate. The frequency (cm^{-1}) of the symmetric and antisymmetric $> SO_2$ stretching bands for each isotopomer are shown below each formula. (*d*) The synthesis of phenyl [$^{16}O,^{17}O,^{18}O$]-sulphate **9**. (*e*) The stereochemical course of sulphuryl transfer catalysed by the aryl sulphotransferase from *Aspergillus oryzae*. Although the configuration of the phenyl [$^{16}O,^{17}O,^{18}O$]sulphate and the *p*-cresyl [$^{16}O,^{17}O,^{18}O$]sulphate have not yet been determined, both gave 1-O-benzoyl-3(*R*)butane-diol-3-[$^{16}O,^{17}O,^{18}O$]-sulphate on chemical transfer to 1-O-benzoyl-3(*R*)butane-diol with the same configuration at sulphur. The enzyme catalysed sulphuryl transfer must therefore have proceeded with retention of configuration at sulphur.

were observed in both the symmetric and antisymmetric $> SO_2$ stretching regions (figure 3). For the isotopomer containing ^{18}O in the C–O–S bridge the symmetric and antisymmetric $> SO_2$ absorption bands were at 1201 and 1414 cm^{-1}, respectively, i.e. identical (at 1 cm^{-1} resolution) with those for $(4R)$-4-methyl-2,2-dioxo-1,3,2-dioxathiane (and consequently not resolved from a small amount of unlabelled material). Thus a heavy oxygen isotope in the C–O–S bridge of the cyclic sulphate ester leaves both the symmetric and antisymmetric $> SO_2$ stretching frequencies unperturbed.

As none of the S–O bonds are broken in the cyclization of $(1R)$-3-hydroxy-1-methylpropyl sulphate with sulphuryl chloride the cyclization should proceed stereospecifically for a chiral $[^{16}O,^{17}O,^{18}O]$sulphate with retention of configuration. In order to confirm this prediction the $[(S)$-$^{15}O,^{17}O,^{18}O]$-sulphate ester **7** and the $[(R)$-$^{16}O,^{17}O,^{18}O]$sulphate ester **8** were cyclized with sulphuryl chloride and the FTIR spectra of the isotopomeric mixture of cyclic sulphate esters measured. The spectra of the symmetric and antisymmetric $> SO_2$ stretching frequencies are shown in figure 4.

Figure 2c shows the mixture of isotopomeric $(4R)$-4-methyl-2,2-dioxo-1,3,2-dioxathianes that should be formed by cyclizing the (S_S)- and (R_S)-chiral $[^{16}O,^{17}O,^{18}O]$sulphate esters **7** and **8** with retention of configuration at sulphur by the mechanism outlined in figure 2b. If all three isotopes were fully enriched only the three isotopomers shown on the top row of each set would be obtained, but in practice the '^{17}O-site' consists of a substantial amount of ^{16}O and ^{18}O, and therefore nine isotopomeric species should be formed; the ^{18}O site is 99% atom ^{18}O. The symmetric and antisymmetric $> SO_2$ stretching frequencies are shown for each isotopomer.

The spectra shown in figure 4a,b are easily distinguishable, and therefore provide a method for the stereochemical analysis of chiral $[^{16}O,^{17}O,^{18}O]$sulphate esters of unknown configuration (Lowe & Parratt 1988).

SYNTHESIS OF PHENYL [^{16}O,^{17}O,^{18}O]SULPHATE

To study both chemical and enzyme catalysed sulphuryl transfer reactions phenyl $[^{16}O,^{17}O,^{18}O]$-sulphate of known absolute configuration was required. The route developed for the synthesis of phenyl $[^{16}O,^{17}O,^{18}O]$sulphate is outlined in figure 2d. $[^{18}O]$Thionyl chloride, prepared by the reaction of $[^{18}O_2]$sulphur dioxide with 1,4-bis(trichloromethyl)-benzene in the presence of a catalytic amount of ferric chloride (Hepburn & Lowe 1989), was reacted first with phenol (one equivalent) and then 3β-cholestanol (one equivalent) to give a single crystalline diastereoisomeric $[^{18}O]$sulphite diester. An X-ray crystal structure analysis of the unlabelled material is in progress which, when complete, should establish the absolute configuration at sulphur. The $[^{18}O]$sulphite diester was oxidized with ruthenium $[^{17}O]$tetroxide (generated by

isotopic exchange between ruthenium tetroxide and $[^{17}O]$water). The $[^{17}O,^{18}O]$sulphate diester on treatment with tetrabutylammonium azide in methylene chloride released phenyl $[^{16}O,^{17}O,^{18}O]$sulphate. Since none of the sulphur–oxygen bonds are perturbed in this process, and since the oxidation of sulphite diesters to sulphate diesters occurs with retention of configuration (Lowe & Salamone 1983) the absolute configuration the phenyl $[^{16}O,^{17}O,^{18}O]$-sulphate will be established when the configuration at sulphur in the sulphite diester has been determined.

CHEMICAL AND ENZYME CATALYSED SULPHURYL TRANSFER REACTIONS

The tetrabutylammonium salt of the phenyl $[^{16}O,^{17}O,^{18}O]$sulphate and 1-O-benzoyl-3(R)butane-diol were dissolved in carbon tetrachloride and the solution kept at 100 °C for 16 h in a sealed vessel by which time complete transfer had occurred. After debenzoylation of the 1-O-benzoyl-3(R)butane-diol-3-$[^{16}O,^{17}O,^{18}O]$sulphate and extraction of phenol, cyclization was performed on the pyridinium salt with sulphuryl chloride. From the FTIR analysis of the purified isotopomers of $(4R)$-4-methyl-2,2-dioxo-1,3,2-dioxathiane it was clear that the 1-O-benzoyl-3(R)-butane-diol-3-$[^{16}O,^{17}O,^{18}O]$sulphate had the (S_S) configuration. Although the stereochemical course of the chemical transfer will not be known until the

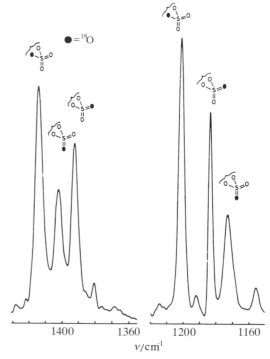

Figure 3. The FTIR spectrum showing the symmetric and antisymmetric $> SO_2$ stretching frequencies of the isotopomeric mixture obtained by cyclizing $(1R)$-3-hydroxyl-1-methylpropyl $[^{18}O]$sulphate with sulphuryl chloride. The symmetric and antisymmetric SO_2 stretching frequencies at 1201 and 1414 cm^{-1}, respectively coincide with those for unlabelled $(4R)$-4-methyl-2,2-dioxo-1,3,2-dioxathiane. Only partial structures, showing the isotopic arrangement around sulphur are shown.

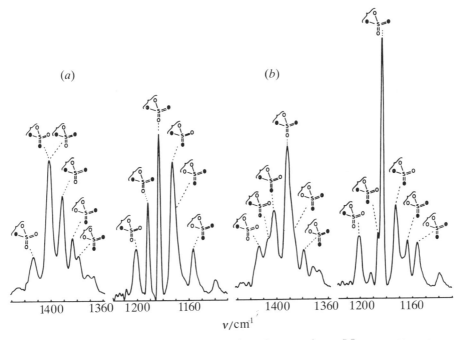

Figure 4. The FTIR spectra showing the symmetric and antisymmetric $>SO_2$ stretching frequencies of the isotopomeric mixture of 4-methyl-2,2-dioxo-1,3,2-dioxathianes obtained by cyclizing with sulphuryl chloride: (a) the [(S)-$^{16}O,^{17}O,^{18}O$]sulphate 7 in which the '^{17}O-site' consists of 37.4 atom % ^{16}O, 36.4 atom % ^{17}O, and 26.2 atom % ^{18}O; (b) the [(R)-$^{16}O,^{17}O,^{18}O$]sulphate 8 in which the '^{17}O-site' consists of 36.0 atom % ^{16}O, 37.1 atom % ^{17}O, and 26.9 atom % ^{18}O. Only partial structures, showing the isotopic arrangement around sulphur, are shown.

configuration of the phenyl [$^{16}O,^{17}O,^{18}O$]sulphate has been established this result allows the stereochemical investigation of enzymic reactions to be determined.

Phenyl [$^{16}O,^{17}O,^{18}O$]sulphate was now incubated with a two molar equivalents of *p*-cresol in the presence of the aryl sulphotransferase from *Aspergillus oryzae* (figure 2 *e*). This enzyme was originally considered to be an aryl sulphatase until it was realized that the rate of transfer was greatly increased if a phenolic acceptor was available (Burns & Wynne 1975). Thus in the presence of a phenolic acceptor the sulphatase activity is suppressed. Interestingly, the enzyme was shown not to require adenosine 3′,5′ diphosphate as a co-factor but a kinetic study suggested the involvement of a covalent sulpho-enzyme intermediate, the transfer of the sulphuryl group from donor to acceptor occurring by a ping-pong type mechanism (Burns *et al.* 1977).

The [$^{16}O,^{17}O,^{18}O$]sulphuryl transfer was allowed to proceed only until 10% of the phenyl [$^{16}O,^{17}O,^{18}O$]-sulphate had been depleted, since if the reaction proceeds with inversion of configuration and the product *p*-cresyl [$^{16}O,^{17}O,^{18}O$]sulphate could act as a donor, partial or even total racemization would occur. The phenyl- and *p*-cresyl [$^{16}O,^{17}O,^{18}O$]sulphates were separated by HPLC and the configuration of the *p*-cresyl [$^{16}O,^{17}O,^{18}O$]sulphate analysed by transferring the [$^{16}O,^{17}O,^{18}O$]sulphate to 1-O-benzoyl-3(R)butane-diol in carbon tetrachloride at 100 °C for 16 h as above. After debenzoylation of the 1-O-benzoyl-3(R)butane-diol-3-[$^{16}O,^{17}O,^{18}O$]sulphate followed by cyclization with sulphuryl chloride, FTIR analysis showed that the spectrum was essentially identical with that derived from the phenyl [$^{16}O,^{17}O,^{18}O$]sulphate. Clearly, the *p*-cresyl

[$^{16}O,^{17}O,^{18}O$]sulphate had the same configuration at sulphur as the phenyl [$^{16}O,^{17}O,^{18}O$]sulphate and hence the enzymic reaction proceeds with retention of configuration at sulphur. Thus, in this first investigation of the stereochemical course of a sulphotransferase, it is gratifying to find that the kinetic and stereochemical methods are in accord, sulphuryl transfer occurring by a ping pong mechanism with a sulpho-enzyme intermediate on the reaction pathway.

I acknowledge the contribution of Roy Bicknell, Paul M. Cullis, Richard L. Jarvest, Salvatore J. Salamone, Martin J. Parratt, Timothy W. Hepburn and Christina L. L. Chai whose published (cited below) and unpublished work are reported in this article. Support of this work by the SERC is also gratefully acknowledged.

REFERENCES

Akagi, J. M. & Campbell, L. L. 1962 *J. Bacteriol.* **84**, 1194.
Baddiley, J., Buchanan, J. G. & Letters, R. 1957 *Proc. Chem. Soc. Lond.* 147.
Baumann, E. 1876 *Arch. ges. Physiol.* **12**, 69; **13**, 285.
Baumann, E. 1876 *Ber. Dtsch. Chem. Ges.* **9**, 54.
Bernstein, S. & Solomon, S. 1970 *The chemistry and biochemistry of steroid conjugates.* Berlin and New York: Springer–Verlag.
Bicknell, R., Cullis, P. M., Jarvest, R. L. & Lowe, G. 1982 *J. biol. Chem.* **257**, 8922.
Burns, G. R. J., Galanopoulou, E. & Wynne, C. H. 1977 *Biochem. J.* **167**, 223.
Burns, G. R. J. & Wynne, C. H. 1975 *Biochem. J.* **149**, 697.
Farley, J. R., Cryns, D. F., Yang, Y. H. J. & Segel, I. H. 1976 *J. biol. Chem.* **251**, 4389–4397.
Farley, J. R., Nakayama, G., Cryns, D. & Segel, I. H. 1978 *Archs biochem. Biophys.* **185**, 376–390.

Gwynn, R. W., Webster, L. T. Jr & Veech, R. L. 1974 *J. biol. Chem.* **249**, 3248.

Hansen, P. E. 1983 *Ann. rep. Spectrosc.* **15**, 105.

Hepburn, T. W. & Lowe, G. 1990 *J. Labelled Cpds. Radiopharms.* **28**, 617.

Kjaer, A. 1960 *Fortschr. Chem. org. NatStoffe.* **18**, 122.

Kjaer, A. 1961 *Organic sulphur compounds* (ed. N. Kharasch), vol. 1, p. 409. Pergamon Press.

Kirby, A. J. 1983 *The anomeric effect and related stereoelectronic effects at oxygen.* Berlin, Heidelberg and New York: Springer–Verlag.

Lichtenberger, J. 1948 *Bull. Soc. chim. Fr.* 1002.

Lichtenberger, J. & Durr, L. 1956 *Bull. Soc. chim. Fr.* 664.

Lillis, V., Dodgson, K. S., White, G. F. & Payne, W. J. 1983 *Appl. environ. Microbiol.* **46**, 988.

Lowe, G. 1983 *Acct. chem. Res.* **16**, 244.

Lowe, G. & Parratt, M. J. 1988 *Bioorg. Chem.* **16**, 283.

Lowe, G. & Salamone, S. J. 1983 *J. chem. Soc. chem. Commun.* 1392.

Lowe, G. & Salamone, S. J. 1984 *J. chem. Soc. chem. Commun.* 466.

Pinchas, S. & Laulicht, I. 1971 *Infrared spectra of labelled compounds*, pp. 238–280. London and New York: Academic Press.

Pflugrath, J. W. & Quiocho, F. A. 1988 *J. molec. Biol.* **200**, 163.

Risley, J. S. & Van Etten, R. L. 1979 *J. Am. chem. Soc.* **101**, 252.

Robbins, P. W. & Lipmann, F. 1956 *J. Am. chem. Soc.* **78**, 2652.

Robbins, P. W. & Lipmann, F. 1957 *J. biol. Chem.* **229**, 837.

Robbins, P. W. & Lipmann, F. 1958 *J. biol. Chem.* **233**, 686.

Rydel, T. J., Ravichandran, K. G., Tulinsky, A., Bode, W., Huber, R., Roitsch, C. & Fenton II, J. W. 1990 *Science, Wash.* **249**, 277.

Schubert, M. & Hamerman, D. 1968 *A primer in connective tissue biochemistry.* Lea and Febinger.

Seubert, P. A., Hoang, L., Renosto, F. & Segel, I. H. 1983 *Archs Biochem. Biophys.* **225**, 679.

Seubert, P. A., Renosto, F., Knudsen, P. & Segel, I. H. 1985 *Archs Biochem. Biophys.* **240**, 509.

Shapiro, L. J., Cousins, L., Fluharty, A. L., Stevens, R. L. & Kihara, H. 1977 *Pediat. Res.* **11**, 894.

Shoyab, M., Su, L. Y. & Marx, W. 1972 *Biochim. biophys. Acta* **258**, 113–124.

Stadtman, E. R. 1973 The enzymes (3rd edn) (ed. P. D. Boyer), vol. 8, pp. 35–37. New York: Academic Press.

Tweedie, J. W. & Segel, I. H. 1971 *J. biol. Chem.* **246**, 2438–2446.

Vederas, J. C. 1980 *J. Am. chem. Soc.* **102**, 374.

Wilson, L. G. & Bandurski, R. S. 1958 *J. biol. Chem.* **233**, 975–981.

The energetics of intramolecular reactions and enzyme catalysis

MICHAEL I. PAGE

Department of Chemical and Physical Sciences, Huddersfield Polytechnic, Queensgate, Huddersfield HD1 3DH, U.K.

SUMMARY

The relative rates of reactions should always be examined by an awareness of differential effects. The magnitude and variation of the relative rates of intramolecular reactions can be rationalized by the differences in entropy and strain energy. The relative rates of enzyme-catalysed reactions are sometimes due to groundstate effects. The β-lactamase-catalysed hydrolysis of β-lactam antibiotics my require a unique disposition of catalytic groups owing to an unusual process of bond fission in the four membered ring.

1. INTRODUCTION

The beauty of chemistry and biology is that both are largely concerned with the study of change. Even the most sophisticated attempts to investigate an isolated and apparently static state usually involve a perturbation of that state and cause a change in the system. This is the dilemma of the philosophers, so what hope is there for understanding the process of the changes in bonding between atoms when a chemical reaction occurs? Apart from perhaps isotopic substitution, most studies of reactions involve a comparison between different systems and a significant perturbation of the system under investigation. We change the solvent, the temperature, the salt concentration, the substituents in both the substrate and the enzyme and then study the change these perturbations cause on some parameter involving a measurement of the change of state, such as a rate or equilibrium constant. It is not surprising therefore that in all these changes we often lose sight of the fundamental differences between the various states. Relative rate and equilibrium constants can be the result of changes in the energies of the initial state and the transition state or product state, respectively.

2. THE DIFFERENCE BETWEEN INTRA- AND INTERMOLECULAR REACTIONS

The first step in an enzyme-catalysed reaction is the bringing together of the substrate and enzyme. This binding process brings, in most cases, the reacting groups on the enzyme and substrate into close proximity (scheme 1). This approximation of reactants changes their molar free energy and has long been thought to be an important contribution to the efficiency of enzyme catalysis. An analogy is thought to be the efficiency of intramolecular reactions, in which the reactants are covalently linked together, compared with an intermolecular reaction between similar reactant molecules (scheme 1) (Page 1973; Capon & McManus 1976; Kirby 1980; Illuminati & Mandolini 1981; Mandolini 1986).

Even when nucleophilicities, electrophilicities, acidities and basicities etc. are the same in both the intra- and the intermolecular reaction, the former invariably occurs with a greater rate or equilibrium constant. Typical rate enhancements and favourable equilibria of intramolecular reactions are shown in scheme 2 (Capon & McManus 1976). All of these reactions involve the formation of five membered rings either in the product or transition state. However, a comparison of the equilibrium or rate constants for these reactions compared with analogous intermolecular reactions gives ratios varying from 0.5 to 2×10^{12} mol dm^{-3}. The rate enhancements and favourable equilibria have units of concentration because a unimolecular reaction is being compared with a bimolecular one. For this reason the rate enhancement is sometimes called the 'effective concentration' or 'effective molarity', which is the hypothetical concentration of one of the reactants in the intermolecular reaction required to make the intermolecular reaction proceed at the same rate or to the same extent as the intramolecular one. Immediate questions about these effective molarities are why the difference between intra- and intermolecular reactions and why the difference between the relative efficiencies of intramolecular reactions? There has never been a shortage of answers put forward to explain these phenomena and to extrapolate them to theories of enzymic catalysis (see table 1).

Some of these special explanations of the efficiency of intramolecular reactions assume that it is a *rate* phenomenon. However, the variation in the effectiveness of intramolecular reactions is shown by both equilibrium and rate constants (scheme 2), and furthermore, there is often a linear relation between

Phil. Trans. R. Soc. Lond. B (1991) **332**, 149–156
Printed in Great Britain

[43]

149

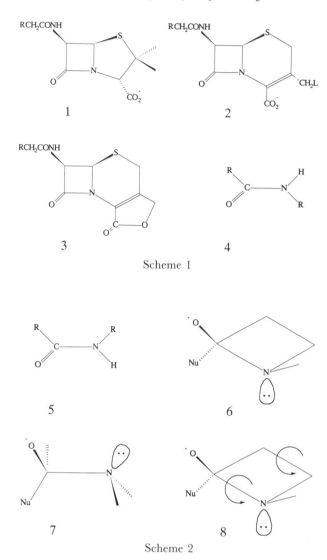

Scheme 1

Scheme 2

Table 1. *Examples of explanations given for the efficiency of intramolecular reactions*

explanation	reference
entropy	Page & Jencks (1971)
approximation, propinquity, proximity	Jencks (1969); Bruice & Benkovic (1965); Koshland (1962)
togetherness	Jencks & Page (1972, 1974)
rotamer distribution	Bruice (1970)
anchimeric assistance	Winstein *et al.* (1953)
distance distribution function	De Lisi & Crothers (1973)
orbital steering	Storm & Koshland (1970, 1972); Dafforn & Koshland (1971)
stereopopulation control	Milstein & Cohen (1970)
substrate anchoring	Reuben (1971)
vibrational activation	Firestone & Christensen (1973)
vibrational activation entropy	Cook & Mckenna (1974)
orbital perturbation theory	Ferreira & Gomes (1980)
group transfer hydration	Low & Somero (1975)
electrostatic stabilization	Warshel (1978)
electric field effect	Hol *et al.* (1979); Van Duijnen *et al.* (1979)
catalytic configurations	Henderson & Wang (1972)
directed proton transfer	Wang (1970)
coupling between conformational fluctuations	Olavarria (1982)
gas phase analogy	Dewar & Storch (1985)
torsional strain	Mock (1976)
circe effect	Jencks (1975)
spatiotemporal postulate	Menger & Venkataram (1985); Menger (1985)
FARCE (freezing at the reactive centres of enzymes)	Nowak & Mildran (1972)

these constants (Page 1977). Attempts to explain the fast rates of intramolecular reactions by critical distances, critical angles, reaction windows, enhanced methods of energy transfer and other phenomena concerned with the rates of reactions are therefore not addressing the full problem.

Changes in equilibrium constants may be understood by examining the variations in the free-energy difference between reactants and products. This thermodynamic approach is enhanced by a knowledge of the nature of reactants and products so that enthalpy and entropy differences can be related to structure. A similar approach may be used for the rates of reactions if the transition state theory of reaction rates is

accepted. Although the structure of the transition state is not known directly, transition state theory allows the application of thermodynamics to understanding the relative rates of reactions.

Large effective molarities are usually the result of a ground or reactant state effect simply because the molar free energy of the intramolecular reactant is greater than that of the intermolecular reactants owing to an entropy difference.

Bringing two molecules together is accompanied by a negative change in entropy because of the reduced volume of space available to the reactants. Mechanically, the increase in order in this process is expressed

$$A+B \rightleftharpoons A\cdots B \rightleftharpoons A-B$$

	loose T.S. or product	tight T.S. or product
$\Delta S/(\text{J K}^{-1} \text{mol}^{-1})$	-40	-150
unfavourable rate or equilibrium factor	10^2	10^8

Scheme 3

mainly as a loss of translational and rotational entropy. The more severely the reactant molecules are confined when brought together the greater is the loss of entropy (Page & Jencks 1971; Page 1973). For most molecules of average shape and size this entropy change makes both the rates and equilibria of bimolecular reactions unfavourable by factors of up to 10^8 M at a standard state of 1 M and at 25 °C (scheme 3).

These changes do not occur in unimolecular reactions and 10^8 M is therefore the maximum difference between an intra- and intermolecular reaction based only on the entropy difference between them and in the absence of strain and solvation effects (Page & Jencks 1971; Page 1973). These calculated entropy changes will not necessarily be reflected in the observed entropies of activation and reaction because the experimental values are dominated by solvent effects especially in water and other polar solvents (Page & Jencks 1971; Page 1973, 1977).

Effective concentrations greater than 10^8 M are the result of strain-energy differences between the two systems being compared. Either the intermolecular reaction shows an unfavourable change in strain energy or the intramolecular one exhibits a release of strain upon ring closure (Page 1973, 1977). It has been shown that there is a good relation between rates and equilibrium constants for intramolecular reactions and the strain-energy changes accompanying ring closure (Page 1977). There is therefore nothing special about these very high effective concentrations and their relevance to enzymic catalysis is limited. Enzymes do not generally owe their efficiency to their ability to induce geometrical strain into their substrates (Page 1984).

The reasons why effective molarities vary, even for reactions involving formation of the same ring size, and are sometimes small, are as follows (Page & Jencks 1971; Page 1973, 1977): (i) unfavourable potential energy changes accompanying the intramolecular reaction, i.e. strain energy is introduced upon ring closure; (ii) unfavourable negative entropy changes in the intramolecular reaction resulting from the loss of internal rotation and a small loss of overall rotational entropy upon ring closure; the loss of internal rotation corresponds to a factor of only about 5–10 per internal rotation, and (iii) favourable positive entropy changes in the intermolecular reaction, resulting from a loose transition state, i.e. a weakly defined geometrical relation between the reactants. The entropy associated with the low frequency vibrations of such flexible transition states compensates for the large negative loss of entropy associated with translation and rotation to give a smaller unfavourable entropy change for bimolecular reactions (scheme 3).

In a series of recent articles it has been suggested that the rates of intramolecular reactions are very dependent upon the distance between the reacting atoms and the time spent at the 'critical distance' at which reaction occurs (Menger 1985; Menger & Glass 1980; Menger & Venkataram 1985). It is claimed this 'spatiotemporal' hypothesis shows that conventional entropy and strain arguments cannot explain the observed magnitude and variation in effective molarities

of intramolecular reactions. However, these descriptions of reactions take no account of the fact that the efficiency of intramolecular reactions is not solely a rate phenomenon – it is reflected also in favourable equilibrium constants. Furthermore, 'the time that reactant molecules reside within a critical distance' is reflected by the entropy of the system. The 'spatiotemporal hypothesis' is a euphemism for entropy. The relative rates of the intramolecular reactions used to justify the spatiotemporal hypothesis may be explained by differences in strain energy and entropy (Page & Jencks 1987).

3. THE DIFFERENCE BETWEEN ENZYME- AND NON-ENZYME-CATALYSED REACTIONS

Because enzymes increase the rate of reactions and show discrimination between possible substrates there has been a temptation to treat these two phenomena separately. Classically the rate enhancement is often attributed to the chemical mechanism used by the enzyme to bring about transformation of the substrate. The fidelity of enzyme and substrate is accounted for by binding – as in the analogy of the 'lock and key'. It is now apparent that these simple ideas need to be reappraised. Chemical catalysis alone cannot explain the efficiency of enzymes. The forces of interaction between the non-reacting parts of the substrate and enzyme may also contribute to a lowering of the activation energy of the reaction (Page 1990). Catalytic groups of equal nucleophilicity and electrophilicity, etc. to those used by the enzyme cannot increase the rate of reactions to even a minute fraction of that achieved by enzymes in the absence of the rest of the protein structure.

The binding energy between substrate and enzyme may be used in a variety of molecular mechanisms to lower the activation energy of the reaction such as charge neutralization, desolvation, geometrical and entropy effects.

It is now generally considered that maximum binding energy, i.e. stabilization, occurs between the substrate and enzyme in the transition state of the reaction. There is an exception to this generalization for the case in which an enzyme equally stabilizes the groundstate and transition state, but catalysis can only occur in this situation if the enzyme is working below but not above saturation. There is a limit to this type of catalysis because if the enzyme binds the transition state and groundstate very tightly, the concentrations of enzyme and substrate required to maintain non-saturation conditions will be decreased. These concentrations could be so low that enzyme catalysis may not be observed. If a non-reacting substituent of a specific substrate contributes a large amount of binding energy, it is essential that this is not expressed in the groundstate or intermediate states in order to avoid saturation conditions and the low concentrations of enzyme and substrate required to observe non-saturation.

Maximum catalytic efficiency may be achieved by

Phil. Trans. R. Soc. Lond. B (1991)

the enzyme stabilizing all transition states, but not intermediate states, in the pathway between reactants and products. It is obvious that an efficient enzyme must stabilize the transition state(s) of a reaction, but it is equally important that the enzyme does not excessively stabilize any intermediate. Stable enzyme intermediate states will bring about saturation conditions at a low concentration of substrate and valuable enzyme will be then tied up in an energy well (Jencks 1975; Page 1984; Knowles 1987; Burbaum *et al.* 1989).

For a given amount of binding energy between the substrate and enzyme the most effective catalysis will be obtained if this energy is used to stabilize the transition state, which maximizes the value of k_{cat}/K_m. For a given free energy of activation, and for a given substrate concentration, the maximum rate is obtained if the substrate is bound weakly, i.e. shows a high K_m. A low value of K_m, i.e. strong binding of the substrate or intermediate state, mediates against catalysis. In agreement with these ideas, the physiological concentrations of most substrates are below their K_m values (Fersht 1974).

A major question in understanding the discrimination between substrates is how does the system allow the binding energy between the 'non-reacting' parts of the enzyme and substrate to be expressed in the transition state and not the groundstate? If a 'nonspecific' and a 'specific' substrate have identical reaction centres, the intrinsic binding energy between the reacting groups undergoing electron density changes and the enzyme should be similar. Differential binding of the transition states around the reaction centre is easy to visualize. Recognition of the geometrical and electronic differences between groundstates and transition states depends on suitably placed amino acids in the enzyme. How can the binding energy of the non-reacting part of the specific substrate stabilize the transition state but not overstabilize the enzyme–substrate or enzyme–intermediate complex? This is very relevant to building in recognition sites in enzyme inhibitors because it is often the binding energy between the non-reacting part of the substrate that accounts for catalysis. The binding energy could be used to 'destabilize' the substrate or to compensate for thermodynamically unfavourable processes necessary for reaction to occur. The importance of this interaction is shown by the observation that the observed binding constant (usually K_m values) of a specific substrate is often apparently 'weaker' or no 'tighter' than that for a non-specific substrate.

The preceding statements should not be interpreted too literally. That 'the enzyme stabilizes the transition state and destabilizes the groundstate' should not encourage us to believe that enzymes are there in a fixed state doing 'something' to the substrate. Stabilization is a mutual process – one could equally well describe the substrate in the transition state stabilizing the enzyme. The important change is the *difference* in energy between the groundstate – enzyme and substrate in their own environment – and the transition state of the enzyme–substrate complex. For example, the statement that 'a positive charge on the enzyme stabilizes the negative charge on the substrate' could

easily be inverted. Furthermore, a positive charge on an enzyme would probably be 'neutralized' in some way in the groundstate of the isolated enzyme, for example, by anionic charges on a neighbouring part of the protein chain or buffers, and by solvation provided by both the rest of the enzyme and water. The rate of reaction would be controlled by the *difference* in free energy provided by these stabilizations in the groundstate and the transition state.

Although enzymes are often considered to be anthropomorphic and have found out how 'to get something for nothing', this is a doubtful achievement even for evolutionarily perfect enzymes. Enzymes cannot solvate and distort substrates with no energetic loss to themselves. The molecular recognition between enzymes and ligands depends on the *difference* between interactions of the isolated molecules and their environment, and the binding-energy interactions between the enzyme and ligand in the complex. Catalysis by enzymes depends on the *difference* between interactions in the transition state and those in the initial state of the isolated molecules (Jencks 1975; Page 1984, 1988). Very rarely will an interaction take place in one state that is completely absent in the other.

The interaction between the substrate S and its environment X (which is usually the solvent) has a complementary interaction in the enzyme E and its environment Y (which may also be the solvent or another site within the enzyme). The changes in interactions that occur when a complex is formed between E and S include those generated in the pair X and Y

$$S \cdots X + E \cdots Y \rightleftharpoons E \cdots S + X \cdots Y. \qquad (1)$$

The net change in the overall energetics depends on the difference in the free energy resulting from this interchange of the 'bonding' partners. The free-energy change reflects the difference in enthalpy and entropy of these interactions. For example, S may have a proton donor site that forms a hydrogen bond with the solvent water acting as an acceptor, X, which is replaced with a similar acceptor site in the enzyme. Usually, in the initial state, this same site in the enzyme will be hydrogen-bonded to water, and $X \cdots Y$ of equation (1) would correspond to the formation of a new hydrogen bond between water molecules, which accompanies enzyme–substrate complex formation (Page 1988; Fersht 1985).

At the reaction centre, electron density and geometrical changes in the substrate, on going to the transition state, may be stabilized by complementary charges and shape of the active site in the enzyme. These electron-density changes could be accompanied by a conformational change so that a large nonreacting group not bound to the enzyme in the groundstate becomes bound in the transition state. In general, however, it is less easy to see how the binding energy of the non-reactive part of the substrate is prevented from being fully expressed in the ES complex unless it is used to compensate for unfavourable processes. If the environment provided by the enzyme is unfavourable to the groundstate structure of the substrate, the binding energy from the non-reacting part of the substrate could be used to force the substrate into this 'unwelcoming hole'. Examples are: (i) a 'rigid' enzyme that has an active site complementary in shape to the transition state but not the groundstate; (ii) an active site that is non-polar and conducive to stabilizing a neutral transition state but destabilizing a charged substrate; (iii) desolvation or solvation changes of groups on the substrate or enzyme; and (iv) an active site where electrostatic charges on substrate and enzyme are similar.

If the binding energy is used to compensate for the induction of 'strain' in the substrate, it is essential that the 'strain' is relieved in the transition state to increase the rate of reaction. This would be the case if the changes in geometry of the substrate were in the direction that accompanies the reaction mechanism. The observed binding energy would be what is 'left over' after the strain has been 'paid for' and may thus appear to be weak. It is essential that the non-reacting group only exhibits its binding energy when the reactive centre of the substrate is bound to the active site. Alternative binding modes would otherwise result, leading to non-productive binding that does not affect k_{cat}/K_m and specificity, but does decrease K_m, leading to saturation conditions at lower concentrations of substrate. The binding energy of the substituent of the specific substrate may be used to prevent non-productive binding (Fersht 1985).

Amino acid side chains, not necessarily those near the active site, may have their exposure to water changed during catalysis. For example, a polar group may become exposed or protected from water on going from the groundstate to transition state, which would make a favourable and unfavourable contribution to the free energy of activation, respectively (Page 1984).

Probably the most important way that the binding energy is 'used' is to compensate for the unfavourable entropy change that accompanies formation of the ES and ES‡ complex (Page & Jencks 1971; Page 1977). The entropy loss that is required to reach the transition state may already have been partially or completely lost in the ES complex.

At 25 °C for a standard state of 1 M, the complete restriction of medium-sized substrates requires a decrease in entropy and an increase in energy of about 150 kJ mol⁻¹, an unfavourable factor of 10⁸. This entropy change is that typically required to form a covalent bond in which the atoms are confined to a relatively small volume because of the loss of translational and rotational freedom (Page & Jencks 1971; Page 1973). If the enzyme-catalysed reaction requires the formation of a covalent bond then the binding energy of the non-reacting part of the substrate may compensate for this necessary but unfavourable entropy change. A non-specific substrate may have insufficient binding energy to compensate for the required entropy loss, resulting in a reduced value of k_{cat}/K_m. If the chemical mechanism of catalysis requires the involvement of other functional groups such as general acids or bases, metal ions or a change in solvation, then a further entropic advantage may be apparent. However, the contribution is smaller than that from covalent catalysis because the 'flexibility' of

hydrogen bonds and metal ion coordination is greater than that for covalent bonds (Page 1984).

4. THE DIFFERENCE IN RATES DUE TO SOLVENT CHANGES

The use of binding energy to compensate for unfavourable solvation changes necessary for chemical reaction is a possible mechanism of catalysis because of the large solvation energies of groups and ions in water (Page 1984). Lone pairs which may act as general bases or nucleophiles will usually be 'solvated' by hydrogen bonding from either water or intramolecularly from the enzyme. Similarly, other potentially reactive groups will usually be 'charge neutralized' in the initial state of the enzyme. These reactive groups will normally require 'desolvation' before bond making or breaking can occur. This process is energetically expensive and yet an essential part of the normal activation energy, but may be compensated by favourable interactions between the substrate and enzyme. Attempts to elucidate the importance of solvation have traditionally been investigated by determining the effect of changing the solvent on the reaction rate. Again it is essential to appreciate that any changes in rate result from the *differences* in groundstate and transition state energies which the solvent may cause.

The parameter k_{cat}/K_m measures the free-energy difference between transition state ES‡ and groundstate E and S (equation 2). If γ_S, γ_E and $\gamma^‡$ represent the activity coefficients of the substrate, enzyme and transition state, respectively, and are defined relative to a common value of unity in purely aqueous solution then k_{cat}/K_m for the enzyme-catalysed reaction in a given solvent mixture is related to its value k_{cat}^E/K_m^o, in pure aqueous solution by equation (3). For sparingly soluble substrates γ_S may be obtained from the solubilities, S and S^o, measured respectively in the presence and absence of organic solvent (equation 4). The major difficulty is to estimate solvation effects on the enzyme because of the lack of a method for measuring γ_E. This can be overcome if solubility and kinetic data are obtained for two substrates S_1 and S_2. In a given solvent system it then becomes possible to eliminate γ_E because this is the same for both substrates. The effect of the solvent on the ratio of the two transition-state activity coefficients is given by equation (5). The right-hand side of equation (5) contains only measurable quantities and the transition state ratio for the two substrates S_1 and S_2, $\gamma_1^‡/\gamma_2^‡$ may be compared with the groundstate ratio γ_1/γ_2,

$$E + S \xrightarrow{k_{cat}/K_m} ES^‡ \tag{2}$$

$$k_{cat}/K_m = (k_{cat}^o/K_m^o)(\gamma_S\gamma_E/\gamma^‡) \tag{3}$$

$$\gamma_S = S^o/S \tag{4}$$

$$\gamma_1^‡/\gamma_2^‡ = \frac{(k_{cat}/K_m)_2 S_2 (k_{cat}^o/K_m^o)_1 S_1^o}{(k_{cat}/K_m)_1 S_1 (k_{cat}^o/K_m^o)_2 S_2^o}. \tag{5}$$

The rate of the α-chymotrypsin-catalysed hydrolysis

of 4-nitrophenyl acetate and N-acetyl-L-trypotophan methyl ester in organic solvent mixtures decreases with increasing amounts of dioxane or propan-2-ol. Measurement of the solubilities of the substrates in the solvent mixtures show that the difference in reactivity of the two substrates with solvent composition is largely a groundstate effect (Bell *et al.* 1974). The organic solvent stabilizes the non-polar substrates relative to water. Changing solvents, or as in the next section, changing substituents, can change the energetics of a reaction by stabilizing or destabilizing the groundstate and these effects can be just as, if not more, important than those in the transition state.

5. THE DIFFERENCE IN RATES CAUSED BY CHANGING SUBSTITUENTS

The effects of changes in substrate structure on enzyme catalytic activity are often used to identify specific binding sites between parts of the enzyme and parts of the substrate. However, changes in reactivity are not necessarily caused by changes in binding interactions between the substrate and the enzyme. Changes in substrate structure can induce different intrinsic 'chemical' effects such as inductive, resonance and steric ones which may relatively stabilize or destabilize the reactant or transition states. Ideally, these effects should be separated from those inter-molecular interactions between the substrate and the enzyme. One method of achieving this is to compare the relative rates of the enzyme-catalysed reaction, brought about by changes in substrate structure, with those of a mechanistically similar non-enzyme-catalysed reaction. It then becomes apparent that some changes in the reactivity of enzyme-catalysed reactions are due to intrinsic changes in substrate reactivity and are not caused by interactions with the enzyme.

β-Lactamases catalyse the hydrolysis of the β-lactam ring of penicillins (1) and other β-lactam antibiotics. The serine β-lactamases are thought to hydrolyse their substrates by the intermediate formation of an acyl-enzyme, similar to the process used by the serine proteases. However, unlike the latter, the nature of the general acid–base groups on the protein involved in proton transfer are not known. The pH-rate profile for k_{cat}/K_m for the serine β-lactamases shows a broad-shaped pH dependence with unknown ionizing groups of pK_a 5.6 and 8.6 (Page & Laws 1990; Buckwell & Page 1987).

β-Lactam antibiotics invariably have an anionic group, such as a carboxylate, at C3 in penicillins (1) or at C4 in cephalosporins (2). It is often assumed that this anionic group binds to lysine-234 in β-lactamase (Buckwell & Page 1987; Page & Laws 1990).

Changing the C4 carboxylate in cephalosporin to the corresponding lactone (3) actually increases the efficiency of the enzyme-catalysed reaction with k_{cat}/K_m increasing 54-fold. It is tempting to interpret this as evidence that the lactone (3) is a 'better' substrate than the cephalosporin (2) because of better recognition or binding. However, the lactone (3) is

130-fold more reactive towards hydroxide ion in its alkaline hydrolysis than is the corresponding cephalosporin (2). The increase k_{cat}/K_m for the lactone is therefore probably an intrinsic chemical effect within the substrate (Laws & Page 1989). However, the rate enhancement brought about by the enzyme is as great for the lactone (3) as it is for the cephalosporin (2). How can this be if the carboxylate group in the antibiotic is a primary recognition site? Either the latter is not the case or the lactone group contributes a similar binding energy to that of the carboxylate.

Interestingly, the pH-rate profile for the β-lacta-mase-catalysed hydrolysis of the lactone (3) shows a similar bell-shape to that for 'normal' substrates with the rate decreasing at high pH owing to an ionizing group of pK_a 8.6. Either this group does not correspond to a cationic one on the protein that binds the carboxylate or its deprotonation decreases the binding energy to both the carboxylate and the lactone groups. The latter explanation is conceivable because k_{cat}/K_m measures the difference between the reactant state structures and the transition state. Although it is reassuring to picture the mutual stabilization of a negatively charged carboxylate anion and a positively charged aminium ion, both groups will be almost certainly equally stabilized in the groundstate by solvent or complementary charged groups. The net binding energy represents the difference between these two states. Taking a carboxylate anion and an aminium ion out of their native environment and placing them close to each other may not therefore be energetically any more favourable than doing the same for a lactone. Although a lactone is not as polarized as a carboxylate anion, its solvation energy in the groundstate is correspondingly reduced. The difference in binding between the two systems may therefore be small. Alternatively, the carboxylate anion and the lactone may both *not* bind strongly to a recognition site on the enzyme, which could explain the difference in β-lactamase reactivities between cephalosporins and penicillins.

6. THE DIFFERENCE IN RATES DUE TO GEOMETRICAL FACTORS IN THE SUBSTRATE

It is of interest to compare the relative reactivity of substrates caused by geometrical differences, particularly those due to changes brought about by cyclization. For example, amides and peptides generally adopt the *trans*-Z-configuration (4) whereas small ring lactams of necessity adopt the *cis*-E-configuration (5). Peptidases and β-lactamases have some properties in common, for example, both groups have serine and zinc enzymes amongst their members (Buckwell & Page 1987; Page & Laws 1990). However, in general, peptidases do not catalyse the hydrolysis of lactams and lactamases do not catalyse the hydrolysis of peptides or amides.

Constraining β-lactams into bicyclic systems such as the penicillins (1) enforces even more differences between lactams and amides. Unlike the normally

Phil. Trans. R. Soc. Lond. B (1991)

planar peptide bond, bicyclic β-lactams are butterfly shaped with the nitrogen being significantly out of the plane defined by its three substituents (Page 1987). The pyramidal configuration of nitrogen thus suggests its lone pair is predominantly on the α-exo and less sterically hindered side of the bridge structure (**1**) (Page 1987). Unlike planar peptides nitrogen inversion and rotation about the carbonyl carbon–nitrogen bond is not possible. It is often suggested that stereoelectronic effects (Deslongchamps 1975) dictate that nucleophilic attack upon amides should occur in such a direction that the nitrogen lone pair becomes antiperiplanar to the newly formed bond. This is unlikely to occur in penicillins as it would involve the nucleophile approaching from the sterically hindered β-endo face of the bicyclic system. In fact, in non-enzyme-catalysed reactions it has been shown that nucleophiles attack from the more favourable endo direction (Page 1987). If this process occurs in the β-lactamase-catalysed reaction the inference is that the configuration of the first formed tetrahedral intermediates are different for β-lactamases and peptidases, (**6**) and (**7**), respectively. Furthermore, as nitrogen inversion and carbon–nitrogen bond rotation cannot take place in (**6**), collapse of the tetrahedral intermediate will be facilitated by protonation on the nitrogen lone pair from the exo-face.

As described in the previous section the nature and location of the general acid–base catalytic groups assumed to be important for the serine β-lactamases are not known. It has, in fact, been suggested (Hertzberg & Moult 1987) that there are none and that the proton from the serine hydroxyl group is transferred directly to the β-lactam nitrogen. Chemical intuition, a dangerous concept, informs us that this is an unlikely process.

Another strange problem associated with β-lactams is that they do not undergo ring opening as readily as would be suggested by the strain energy of the four-membered ring (Page 1987; Webster *et al.* 1990). One possible explanation is that unlike acyclic systems in which carbon–nitrogen bonds are broken by a stretching motion, four-membered rings may be opened by rotation (**8**). A consequence of this would be that a proton donor on the enzyme which facilitates carbon–nitrogen bond cleavage would be situated in a different position than that required in the hydrolysis of acyclic peptides. Furthermore, substituents attached to the incipient amino group would move considerably upon ring opening the β-lactam. It would then be inappropriate to have the substrate's carboxylate tightly bound to the enzyme.

Serine β-lactamase catalyse the hydrolysis of both monocyclic and bicyclic β-lactams but not apparently acyclic amides and anilides. This is consistent with, but of course does not prove, the above hypothesis. It was mentioned in the previous section that binding the carboxylate anion of the β-lactam substrate may not be obligatory. Another relevant observation is that below pH 4 the β-lactamase-catalysed hydrolysis of substrates with carboxylate groups at C3 in penicillins and at C4 in cephalosporins becomes pH independent whereas for the other substituents k_{cat}/K_m carries on decreasing.

This suggests that the undissociated carboxylic acid of the substrate is a good substrate for β-lactamase and is consistent with, but does not prove, the idea that there is not a rigid binding site for the carboxylate anion because this functionality is required to significantly move upon β-lactam ring opening.

REFERENCES

Bell, R. P., Critchlow, J. E. & Page, M. I. 1974 *J. chem. Soc. Perkin Trans.* **2**, 66.

Bruice, T. C. 1970 *The enzymes* (ed. P. D. Boyer), vol. 2, 3rd edn, p. 217. New York: Academic Press.

Bruice, T. C. & Benkovic, S. J. 1965 *Bioorganic mechanisms*, vol. 1, p. 199. New York: W. A. Benjamin Inc.

Buckwell, S. C. & Page, M. I. 1987 β-lactamases – normal peptidases. *Adv. Biosci.* **65**, 24.

Burbaum, J. J., Raines, R. T., Albery, W. J. & Knowles, J. R. 1989 *Biochemistry* **28**, 9293.

Capon, B. & McManus, S. P. 1976 *Neighbouring group participation*. New York: Plenum Press.

Cook, D. B. & McKenna, J. 1974 *J. chem. Soc. Perkin Trans.* 2, 1223.

Danforth, A. & Koshland, D. E. Jr 1971 *Proc. natn. Acad. Sci. U.S.A.* **68**, 2463

De Lisi, C. & Crothers, D. M. 1973 *Biopolymers* **12**, 1689.

Deslongchamps, P. 1975 *Tetrahedron* **31**, 2463.

Dewar, M. J. S. & Storch, D. M. 1985 *Proc. natn. Acad. Sci. U.S.A.* **82**, 2225.

Ferreira, R. & Gomes, M. A. F. 1980 *Proc. 6th Braz. Symp. Theor. Phys.* **2**, 281.

Fersht, A. 1974 *Proc. R. Soc. London.* B **187**, 392.

Fersht, A. 1985 *Enzyme structure and mechanism*. New York: W. H. Freeman.

Firestone, R. A. & Christensen, B. G. 1973 *Tetrahedron Lett.* **389**.

Henderson, R. & Wang, J. H. 1972 *A. Rev. Biophys. Bioengng.* **1**, 1.

Hertzberg, O. & Moult, J. 1987 *Science, Wash.* **236**, 694.

Hol, W. G. J., Van Duijnen, P. T. & Berendsen, H. J. C. 1979 *Nature, Lond.* **273**, 443.

Illuminati, G. & Mandolini, L. 1981 *Acct. chem. Res.* **14**, 95.

Jencks, W. P. 1969 *Catalysis in chemistry and enzymology*. New York: McGraw Hill.

Jencks, W. P. 1975 *Adv. Enzymol.* **43**, 219.

Jencks, W. P. & Page, M. I. 1972 *Proc. Eighth FEBS Meeting, Amsterdam* **29**, 45.

Jencks, W. P. & Page, M. I. 1974 *Biochem. biophys. Res. Commun.* **57**, 887.

Kirby, A. J. 1980 *Advan. phys. org. Chem.* **17**, 183.

Knowles, J. R. 1987 *Science, Wash.* **236**, 1252.

Koshland, D. E. Jr 1962 *J. theor. Biol.* **2**, 75.

Laws, A. P. & Page, M. I. 1989 *J. chem. Soc. Perkin Trans.* 2, 1577–1581.

Low, P. S. & Somero, G. N. 1975 *Proc. natn. Acad. Sci U.S.A.* **72**, 3305.

Mandolini, L. 1986 *Adv. phys. org. Chem.* **22**, 1.

Menger, F. M. 1985 *Acct. chem. Res.* **18**, 128.

Menger, F. M. & Glass, L. E. 1980 *J. Am. chem. Soc.* **102**, 5404.

Menger, F. M. & Venkataram, U. V. 1985 *J. Am. chem. Soc.* **107**, 4706.

Milstein, S. & Cohen, L. A. 1970 *Proc. natn. Acad. Sci. U.S.A.* **67**, 1143

Mock, W. L. 1976 *Bioorg. Chem.* **5**, 403.

Nowak, T. & Mildvan, A. S. 1972 *Biochemistry* **11**, 2813.

Olavarria, J. M. 1982 *J. theor. Biol.* **99**, 21.

Page, M. I. 1973 *Chem. Soc. Rev.* **2**, 295.

Page, M. I. 1977 *Angew. Chem.* **16**, 449

Page, M. I. 1984 *The chemistry of enzyme action* (ed. M. I. Page), pp. 1–54. Amsterdam: Elsevier.

Page, M. I. 1987 *Adv. phys. org. Chem.*, **23**, 165.

Page, M. I. 1988 *J. molec. Cat.* **47**, 241.

Page, M. I. 1990 In *Comprehensive medicinal chemistry*, vol. 2 (ed. P. G. Sammes), pp. 45–60.

Page, M. I. & Jencks, W. P. 1971 *Proc. natn. Acad. Sci. U.S.A.* **68**, 1678.

Page, M. I. & Jencks, W. P. 1987 *Gazz. chim. ital.* **117**, 455.

Page, M. I. & Laws, A. P. 1990 *Molecular mechanisms in bioorganic processes* (ed C. Bleasdale & B. Golding), pp. 319–330. London: Roy. Soc. Chem.

Reuben, J. 1971 *Proc. natn. Acad. Sci. U.S.A.* **68**, 563.

Storm, D. R. & Koshland, D. E. Jr 1970 *Proc. natn. Acad. Sci. U.S.A.* **66**, 445.

Storm, D. R. & Koshland, D. E. Jr 1972 *J. Am. chem. Soc.* **94**, 5805, ibid. **94**, 5817.

Van Duijnen, P. T., Thole, B. T. & Hol, W. G. J. 1979 *Biophys. Chem.* **9**, 273.

Wang, J. H. 1970 *Proc. natn. Acad. Sci. U.S.A.* **66**, 874.

Warshel, A. 1978 *Proc. natn. Acad. Sci. U.S.A.* **75**, 5250.

Webster, P., Ghosez, L. & Page, M. I. 1990 *J. chem. Soc. Perkin Trans.* **2**, 805–813.

Winstein, S., Lindegren, C. R., Marshal, H. & Ingraham, L. L. 1953 *J. Am. chem. Soc.* **75**, 147.

Recent advances in catalytic antibodies

THOMAS S. SCANLON AND PETER G. SCHULTZ

Department of Chemistry, University of California, Berkeley, California 94720, U.S.A.

SUMMARY

Recently the biological machinery of the immune system has been exploited with the aid of mechanistic chemistry to produce catalytic antibodies. Because antibodies can be generated that selectively bind almost any molecule of interest, this new technology offers the potential to tailor-make highly selective catalysts for applications in biology, chemistry and medicine. In addition, catalytic antibodies provide fundamental insight into important aspects of biological catalysis, including the importance of transition-state stabilization, proximity effects, general acid and base catalysts, electrophilic and nucleophilic catalysis, and strain.

1. INTRODUCTION

Chemists have become increasingly sophisticated in their ability to synthesize complex organic molecules and in their understanding of reaction mechanisms. Yet given these spectacular advances, chemistry cannot begin to match the ability of biological systems to synthesize and screen tremendous chemical diversity to produce complex molecules with interesting biological properties. One of the most remarkable examples of this is the immune system. The humoral immune system has the potential to produce on the order of 10^{12} unique receptors. A complex screening mechanism allows the immune system to select from this vast array of molecules, antibodies that bind virtually any synthetic or biomolecule with high affinity and exquisite specificity (Schultz *et al.* 1990). Hybridoma technology, which makes it possible to generate, *in vitro*, large amounts of homogenous antibody molecules, has dramatically expanded the role of antibodies in biology and medicine (Kohler & Milstein 1975). Antibodies are invaluable tools for the detection and analysis of biological materials, are key components of most diagnostic devices and hold promise as highly selective therapeutic and imaging agents.

Recently, we and others have shown that the principles and tools of organic chemistry can be used to exploit the highly evolved machinery of biology, specifically the humoral immune system, to generate a new class of bioactive molecules: catalytic antibodies. By combining the exquisite binding specificity of the immune system with our understanding of chemical and biological reaction mechanisms, antibodies have been generated that catalyse a wide array of chemical reactions from pericyclic rearrangements to peptide bond cleavage. Moreover, several strategies have been developed for generating catalytic antibodies, including the use of antibodies to stabilize negatively charged transition states and to act as entropic traps, and the generation of catalytic groups or cofactor binding sites

in antibody combining sites. Because antibodies can be elicited to a vast array of biopolymers, natural products, or synthetic molecules, catalytic antibodies offer a unique approach for generating enzyme-like catalysts that differentiate complex molecules of biological, chemical and medical interest.

Catalytic antibodies might be used, for example, to develop a family of catalysts analogous to restriction enzymes that cleave proteins or sugars at a particular bond. Such antibodies would be invaluable reagents in biology and might find use as therapeutic agents to selectively hydrolyse protein or carbohydrate coats of viruses, cancer cells, or other physiological targets. Catalytic antibodies also could serve as selective catalysts for the synthesis of pharmaceuticals, fine chemicals, and novel materials. At the same time, the characterization of catalytic antibodies provides fundamental insight into important aspects of biological catalysis, including the importance of transition-state stabilization, proximity effects, general acid and base catalysis, electrophilic and nucleophilic catalysis, and strain.

Several comprehensive reviews on the generation and characterization of catalytic antibodies have been written (Schultz *et al.* 1990; Shokat & Schultz 1990*a*; Schultz 1989). This article reviews recent advances in the field out of our laboratory.

2. TRANSITION STATE STABILIZATION

The first examples of antibody-catalysed reactions were based on the notion of transition-state stabilization. Jencks first proposed over 20 years ago that antibodies raised against a transition-state analogue should selectively bind the transition state of a reaction over substrate and thereby act as a catalyst (Jencks 1969). Enzymes themselves have evolved to provide an active site that is sterically and electronically complementary to the rate-determining transition state. The

Phil. Trans. R. Soc. Lond. B (1991) **332**, 157–164
Printed in Great Britain

157

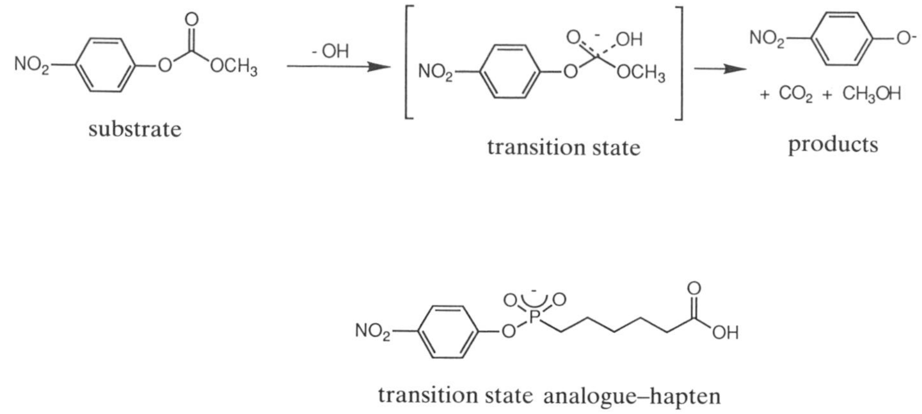

substrate

transition state

products

transition state analogue–hapten

Scheme 1.

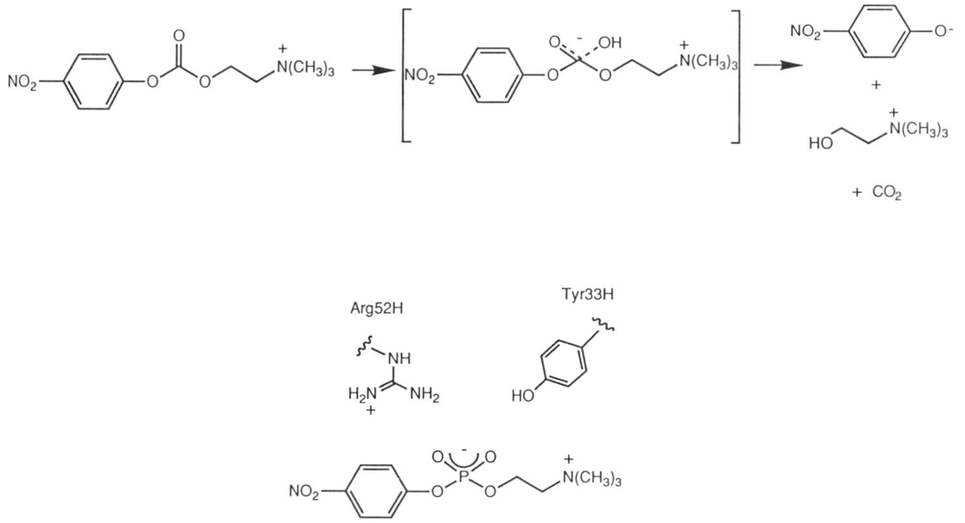

Scheme 2.

first successful application of this notion to the generation of catalytic antibodies was made independently by the groups of Lerner and Schultz in 1986, when they characterized antibodies that catalyse the hydrolysis of esters and carbonates (Tramantano *et al.* 1986; Pollack *et al.* 1986). The rate-limiting step in the uncatalysed reactions is formation of the negatively charged tetrahedral transition state. A stable analogue of such a structure is formed by substitution of a tetrahedral phosphorus atom for the tetrahedral carbon (Bartlett & Marlowe 1983). As predicted, antibodies specific for these transition-state analogues selectively accelerated the hydrolysis of their respective substrates (scheme 1). The antibodies have substantially higher binding affinities for the transition-state analogues than for the reaction substrates, which suggests that they function by stabilizing the transition state. Since these first examples, over 20 acyl transfer reactions have been catalysed by antibodies including ester bond formation, lactone formation and stereospecific ester hydrolysis.

Recently we have used the technique of site-directed mutagenesis to further define the catalytic mechanism and improve the catalytic efficiency of the hydrolytic catalytic antibody S107 (Jackson *et al.* 1991). The

phosphorylcholine (PC) binding antibodies provide a good starting point for investigating antibody binding and catalysis. Two phosphorylcholine binding antibodies, MOPC 167 and T15, have previously been shown to catalyse the hydrolysis of choline carbonates (Pollack *et al.* 1986; Pollack & Schultz 1987). These antibodies belong to a class of highly homologous PC binding antibodies that have been well-characterized with regard to ligand binding kinetics and specificity, biomolecular structure and genetics (Pollet *et al.* 1974; Goetze & Richards 1977; Goetze & Richards 1978; Bennett & Glaudemans 1979; Crews *et al.* 1981; Perlmutter *et al.* 1984). In addition, the three dimensional structure of one representative PC binding antibody, McPC603, has been solved by X-ray crystallography (Satow *et al.* 1986). It has been proposed that T15 and MOPC 167 preferentially stabilize the transition state in carbonate hydrolysis on the basis that phosphorylcholine diesters resemble the tetrahedral negatively charged transition state for the hydrolytic reaction. Consistent with this notion, the antibodies bind the PC transition state analogues with higher affinity than the carbonate substrates. Based on the three-dimensional structures of McPC 603, as well as previous chemical modification and kinetic studies, it has been argued that Tyr 33 and Arg 52 of the heavy chain play a critical role in stabilizing the rate determining transition state configuration via electrostatic and hydrogen bonding interactions. These residues are conserved in all PC specific antibodies (Perlmutter *et al.* 1984). To determine the roles of these residues in binding and catalysis, mutants of the highly homologus PC binding antibody S107 were generated.

Three $Arg\,52_H$ mutants (R52K, R52Q, R52C) and four $Tyr\,33_H$ mutants (Y33H, Y33F, Y33E, Y33D) of wild type S107 were generated by *in vitro* mutagenesis. The mutant and wild type genes were then cloned into a modified SV-2 shuttle vector containing an *Escherichia coli* xanthine-guanine phosphoribosyl transferase (gpt) selection marker (Jackson *et al.* 1991). These constructs were then used to transfect a murine myeloma cell line, and recombinants were selected on azaserine/hypoxanthine. *In vivo* expression in immunosuppressed mice and protein-A affinity chromatography resulted in the isolation of large quantities of pure antibody from ascites fluid.

Mutations at $Arg\,52_H$ which resulted in a loss of the positively charged side chain resulted in a significant decrease in k_{cat}, whereas mutants that retained the positively charged side chain (A52K) had wild type activity. In contrast, mutations at $Tyr\,33_H$ have little effect on catalytic activity. The k_{cat} value of the Y33F mutant as well as the K_m and K_i for carbonate substrate and the phosphodiester transition state analogue, respectively, are comparable to those of wild type antibody. This is somewhat surprising since the X-ray crystal structure (Satow *et al.* 1986) shows the tyrosine hydroxyl group to be within hydrogen bonding distance (2.9 Å†) of the phosphoryl oxygen. Because $Tyr\,33_H$ appears to play no role in binding or catalysis, $Y33H_H$, $Y33E_H$, and $Y33D_H$ mutants were generated in an effort to place a general base in the antibody combining site. The $Y33H_H$ mutant showed a measurable increase in k_{cat} with respect to the wild type antibody, which translated into a 6700 fold overall rate acceleration when compared to the background reaction with 4-methyl imidazole. Although the precise role of the histidine residue in catalysis remains unclear, mechanistic experiments suggest that the histidine may act via a general base mechanism. This mutagenesis study points to the importance of electrostatic stabilization in acyl transfer catalysis by antibodies, and further shows that incremental improvements in antibody catalysis are readily obtained with site-directed mutagenesis.

In addition to negatively charged phosphonates, a number of uncharged transition-state analogues exist for hydrolytic enzymes. The best characterized of these is the potent pepsin inhibitor, pepstatin, which contains the novel amino acid analogue statine. The measured K_i of pepstatin for pepsin is 46 pm, making it one of the most potent enzyme inhibitors known (Rich 1986). Statine can be considered as a 'collected' substrate analogue possessing binding determinants of both the peptide substrate and the enzyme-bound water responsible for addition to the scissile amide

† $1 \text{ Å} = 10^{-10} \text{ m} = 10^{-1} \text{ nm}$.

Scheme 3.

bond. The secondary hydroxyl group is thought to mimic the tetrahedral transition state for amide bond hydrolysis.

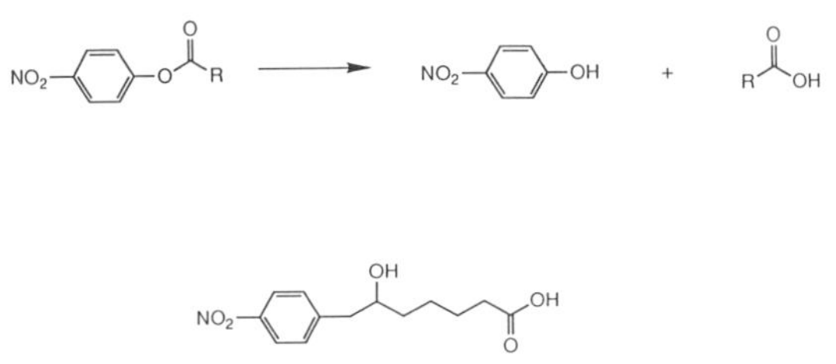

In addition to peptidase inhibitors, uncharged transition-state analogue inhibitors of enzymes in nucleotide biosynthesis pathways have been isolated. Coformycin inhibits the enzyme adenosine deaminase and is one of the most potent inhibitors known for any enzyme with $K_i = 2.5$ pm (Frick *et al.* 1986). The enzyme converts adenosine into inosine, which involves the hydrolysis of the amidine moiety of adenine. Based on the high binding constants of these transition-state analogues, haptens with similar structural features might be expected to elicit catalytic antibodies.

Shokat *et al.* (1991) have recently shown that hydrolytic catalysts can in fact be generated by using the charge neutral statine-based peptidase hapten shown in scheme 3 (Shokat *et al.* 1991). From a panel of five monoclonal antibodies specific for the hapten, one was found that catalysed the hydrolysis of both carbonate and ester substrates. The values of k_{cat} and K_M were 0.72 min^{-1} and 3.65 mm, respectively, for the ester substrate, and 0.31 min^{-1} and 3.33 mm, respectively, for the carbonate substrate. Comparison of K_m/K_i and k_{cat}/k_{uncat} shows that the rate acceleration is between one and two orders of magnitude higher than the expected value based on the differential binding of substrate and hapten-inhibitor. This discrepancy suggests that mechanistic factors which are not clearly understood, and most likely result from the structural diversity of the humoral immune response, are important for a large degree of antibody catalysis in this system. Attempts to generate antibodies that catalyse adenosine deamination by immunizing with a coformycin derivative as a hapten failed to produce any catalytic antibodies.

The notion of transition state stabilization has also recently been applied to the metallation of porphyrins. Metallo-porphyrins represent an important class of biologically relevant cofactors involved in electron and oxygen transport as well as many oxidation reactions. Ferrochelatase, the terminal enzyme in the heme biosynthetic pathway, catalyses the insertion of iron (II) into protoporphyrin (Lavallee 1988). A potent inhibitor of this enzyme is the bent porphyrin, N-methylprotoporphyrin (Dailey & Fleming 1983). The distorted structure of the methylated porphyrin macrocycle results from steric crowding due to the internal methyl substituent, and is thought to resemble the transition state of porphyrin metallation catalysed by ferrochelatase. This distortion of the macrocyclic ring system forces the chelating nitrogen electron pairs into a position which is more accessible for binding of the

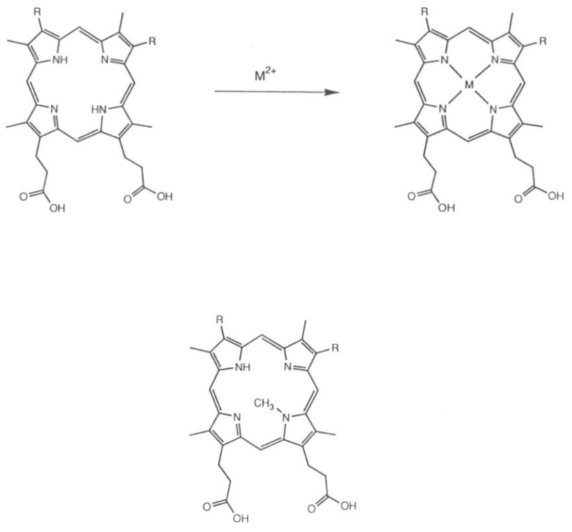

Scheme 4.

incoming metal ion. Cochran & Schultz (1990*a*) have recently shown that antibodies generated against this bent mesoporphyrin efficiently catalyse the metallation of a mesoporphyrin substrate (Cochran & Schultz 1990*a*). Antibody catalysis was observed with a variety of transition metals including Zn (II) and Cu (II). An initial rate analysis was performed for the metallation reaction with Cu (II). At 1 mM Cu (II), the antibody catalysed the metallation reaction with a K_M of 50 μM and an apparent V_{max} of 2.5 μM h^{-1}. Catalysis could be inhibited by free hapten as well as various metalloporphyrins.

The catalytic properties for this antibody are similar to those of the enzyme ferrochelatase. The reported K_M values for ferrochelatase of 10–70 μM (Levallee 1988) are comparable to the value of 50 μM determined for the antibody. Both enzyme and antibody catalysis suffer from product inhibition, and both have comparable affinities for the N-methylporphyrin. Both enzyme and antibody can insert a variety of divalent transition metals into porphyrins, and the turnover numbers (k_{cat}) are similar. A calculated value of k_{cat} for ferrochelatase with Zn (II) is 800 h^{-1}, and the experimentally determined value of k_{cat} for the antibody is 80 h^{-1}. This represents the closest correlation of turnover number achieved to date for a catalytic antibody with the analogous enzyme. In addition, this work shows the use of antibody binding energy to distort substrate conformation in a productive fashion along the reaction coordinate.

3. INTRODUCTION OF CATALYTIC GROUPS

A second important strategy for generating catalytic antibodies involves the rational generation of chemically reactive amino acid side chains in antibody combining sites. Many enzymes utilize a combination of active site residues such as nucleophiles, electrophiles, general acids, and general bases to achieve their remarkable rates. For example, staphylococcal nuclease contains both an active site glutamate residue which functions as a general base to deprotonate water,

in addition to an arginine residue that stabilizes the negatively charged transition state. Shokat & Schultz have shown the viability of introducing a catalytic base into an antibody combining site for an antibody catalysed elimination reaction (Shokat & Schultz 1989). Immunization with a positively charged alkyl ammonium hapten afforded antibodies with a complementary negatively charged carboxylate residue appropriately positioned in the combining site to function as a general base for a β-elimination of hydrogen fluoride. From a panel of six monoclonal antibodies, four were able to catalyse the elimination reaction; one catalysed the reaction with a rate acceleration of approximately 10^5 over the corresponding background reaction. Chemical modification and affinity labelling experiments confirmed the presence of an active site carboxylate, and kinetic isotope effects showed that the rate-determining step involves proton abstraction from the α-carbon atom of the fluoroketone substrate (Shokat & Schultz 1990*b*). In addition, the reaction displays a pH profile pointing to an ionizable active site residue with a pKa of 6.2. This study was the first demonstration that antibody–hapten complimentarity can be exploited to obtain combining site residues capable of participating in chemical transformations.

More recently this approach has been applied to an antibody catalysed *cis–trans* isomerization reaction. Antibodies raised to the bis-nitrophenyl piperidinium hapten shown below were capable of catalysing the isomerization of the corresponding *cis* enone to the *trans* enone (Jackson & Schultz 1991). The catalysis was competitively inhibited by the free hapten, and the antibody accelerated the rate of reaction 15000-fold over the uncatalysed background reaction. A pH dependence study on k_{cat} suggests the presence of an ionizable combining site residue that, with a pKa of 5.5, participates in catalysis.

Additional evidence for a catalytic residue and corresponding covalent antibody–substrate adduct follows from chemical modification experiments including epoxide affinity labelling and treatment of the antibody with diazoacetamide, a reagent that specifically esterifies carboxylate residues.

A mechanistic model consistent with these observations is shown in scheme 6. Initial binding of the *cis* enone isomer to the antibody is followed by 1,4-nucleophilic addition of a carboxylate group leading to a covalent antibody–substrate intermediate. It was anticipated that an active site carboxylate residue

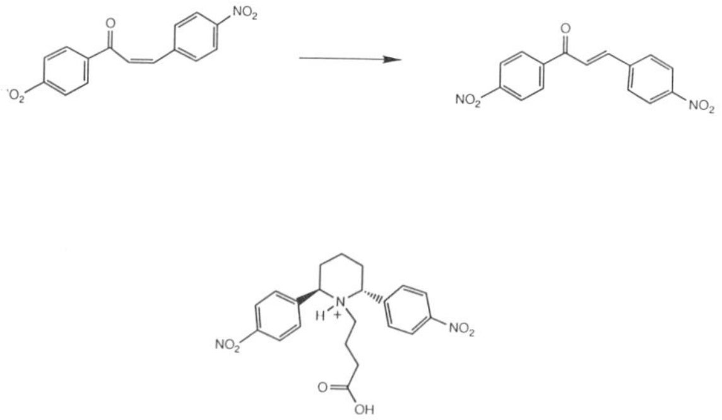

Scheme 5.

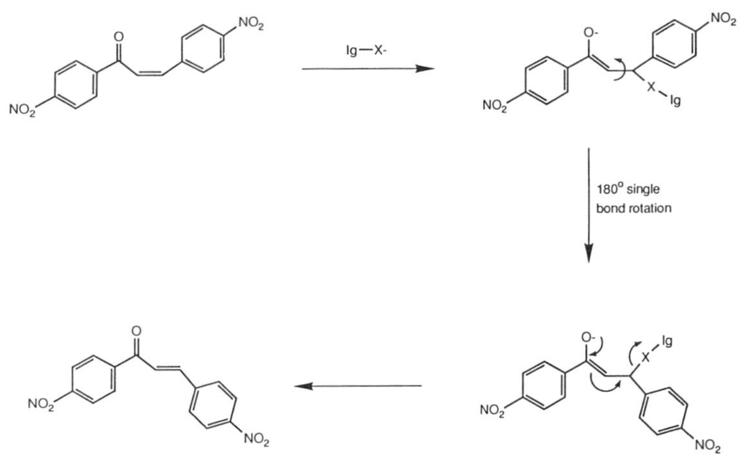

Scheme 6.

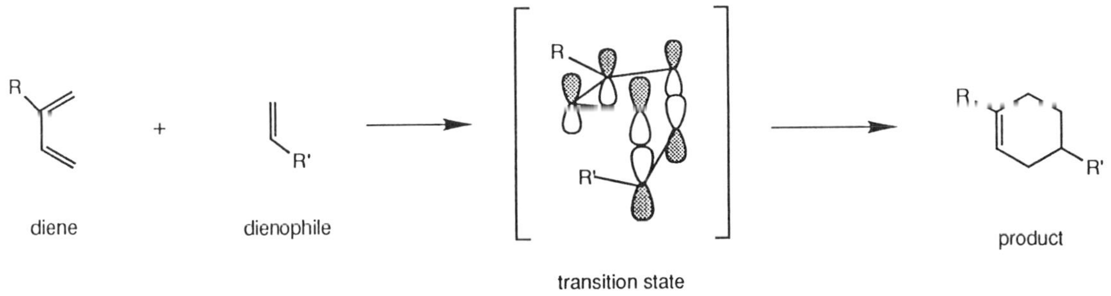

Scheme 7.

might arise in response to the ammonium ion contained in the hapten structure. The nucleophilic 1,4-addition converts the double bond of the substrate into a single bond, about which rotation is facile. Following rotation about this bond, the covalent adduct collapses by elimination, giving rise to the thermodynamically favoured *trans* enone and the regenerated antibody catalyst. For isomerization to occur, the antibody must accommodate the orthogonal transitional state for bond rotation. Computer modelling shows that the *trans* configuration of the nitrophenyl rings in the hapten do in fact mimic the geometry of a transition state in which the α, β- is rotated 90°. Consistent with this observation, only the *trans* and not the *cis* hapten afforded catalytic antibodies.

4. ANTIBODIES AS ENTROPY TRAPS

Another approach toward the design of catalytic antibodies involves the use of antibody binding energy to lower entropic barriers to reactions. Jencks & Page have argued that entropic effects can account for effective molarities of up to 10^8 M in enzyme catalysed reactions (Page & Jencks 1971). These notions have been tested in the design of antibodies that catalyse transacylation reactions (Schultz *et al.* 1990; Shokat & Schultz 1990; Schultz 1989; Napper *et al.* 1987; Janda *et al.* 1988), Diels Alder reactions (Braisted & Schultz 1989; Hilvert *et al.* 1989) and Claisen rearrangements (Jackson *et al.* 1988; Hilvert & Nared 1988). The latter reaction involves the conversion of chorismic acid to prehenic acid. This thermal 3,3-sigmatropic rearrange-

ment occurs through an asymmetric chairlike transition state. One might expect that an antibody combining site that is complementary to the conformationally restricted transition state would accelerate this rearrangement. In fact one antibody elicited to a bicycle transition state inhibitor of chorismate mutase increased the rate of the rearrangement 10000-fold, whereas the enzyme chorismate mutase accelerates the reactions approximately 10^6-fold over the uncatalysed background reaction (Jackson *et al.* 1988). For this antibody, mechanisms involving a cationic substituent effect or general acid catalysis were ruled out. As expected, the entropy of activation of the antibody-catalysed reaction was close to zero, compared with an entropy of activation ($\Delta S^{\ddagger}$) of -13 entropy units for the uncatalysed reaction. This antibody has recently been cloned and introduced into bacteria that lack the ability to synthesize prephenic acid. Random mutagenesis and selection will be used in an effort to increase the catalytic activity of this antibody.

Another example of the use of antibodies to act as entropic traps involves an antibody catalysed Diels–Alder reaction. The Diels–Alder reaction has long been one of the most powerful transformations in organic chemistry. This process consists of a bimolecular reaction between a diene and an alkene giving rise to a cyclohexene product. The transition state involves a highly ordered cyclic array of interacting orbitals in which carbon—carbon bonds are broken and formed in a single concerted mechanistic event (figure 7). Consequently, an unfavourable entropy of activation on the order of -30 to -40 entropy units is generally

Phil. Trans. R. Soc. Lond. B (1991)

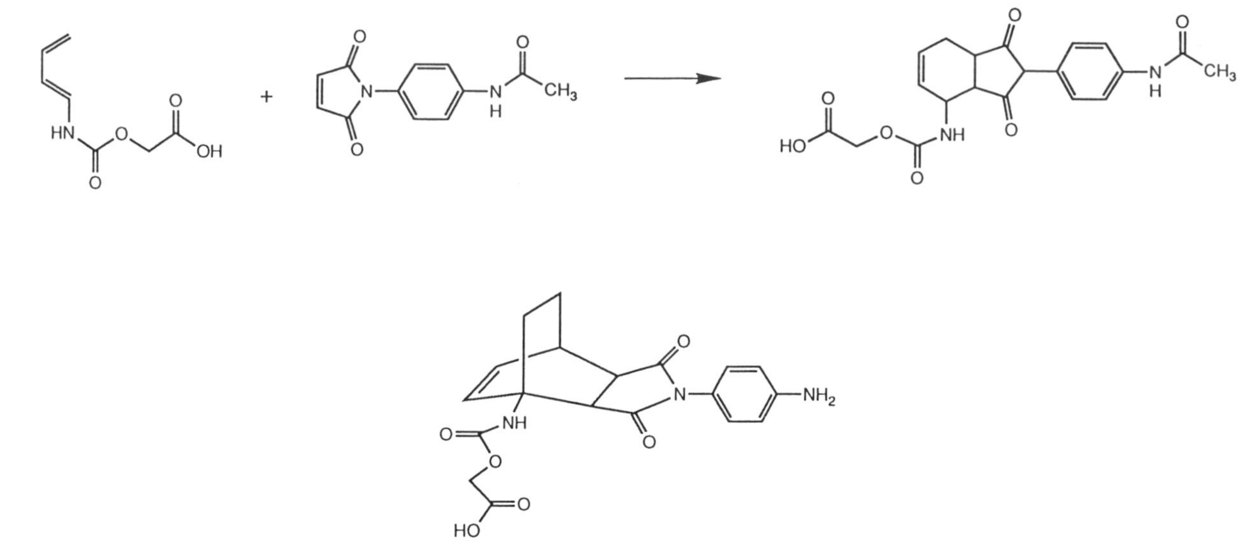

Scheme 8.

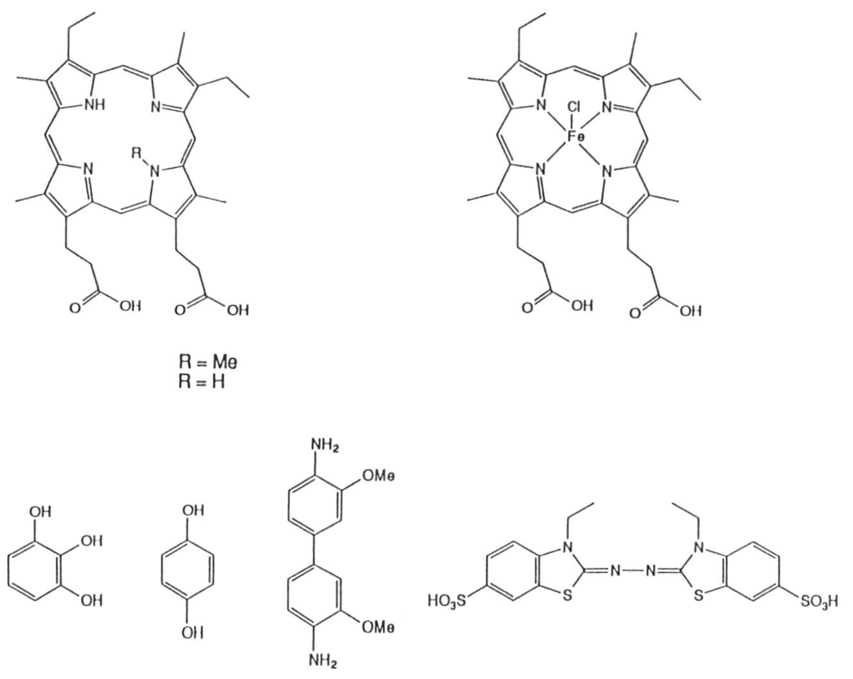

R = Me
R = H

Scheme 9.

observed. The design of a transition state analogue–hapten which would lead to catalytic antibodies for this reaction must address two fundamental issues: (i) an 'entropic sink' must be provided such that the two substrate molecules are oriented in a reactive configuration upon binding, and (ii) a mechanism for eliminating product inhibition must be incorporated, since the cyclohexene product is structurally very similar to the transition state. Hilvert and co-workers were successful in designing a system that satisfies these criteria (Hilvert *et al.* 1989). The particular Diels–Alder reaction chosen was that of tetrachlorothiophene dioxide with N-ethylmaleimide giving rise to an initial tricyclic Diels–Alder adduct which spontaneously extrudes sulphur dioxide, resulting in the dihydrophthalimide product. One of five antibodies raised to a stable [2.2.1] bicyclic transition state analogue was found to significantly accelerate the rate of reaction

over background. The antibody catalysed the reaction with multiple turnovers (> 50), and catalysis could be inhibited by free hapten.

A more general strategy for Diels–Alder catalysis has recently been reported by Braisted & Schultz (1989). The hapten contains an ethano bridge that locks the cyclohexane ring into the conformation resembling the Diels–Alder transition state, and presumably makes hydrophobic binding contacts in the antibody combining site (figure 8). The cyclohexene product does not contain this hydrophobic bridge, and therefore would be bound less tightly than the hapten, and should diffuse from the combining site. The diene substrate in this system is acyclic, and is not covalently locked in the *cis–syn* reactive conformation. Since antibodies are elicited to a locked cyclohexane ring system, the reactive *cis–syn* conformation should be more favourably bound in the combining site.

Phil. Trans. R. Soc. Lond. B (1991)

From a panel of 10 monoclonal antibodies elicited to the hapten, one was found which catalysed the formation of the Diels–Alder adduct with a large rate acceleration over the background reaction. The apparent second order rate constants (k_{cat}/K_m) were found to be 900 $M^{-1} s^{-1}$ for the diene, and 583 $M^{-1} s^{-1}$ for the dienophile. These values can be compared to the second order rate constant for the background reaction of 1.9 $M^{-1} s^{-1}$ in water and 2×10^{-3} $M^{-1} s^{-1}$ in acetonitrile. The reaction is competitively inhibited by free hapten with a K_D of 126 nM, and by product with a K_D of 10 μM. This work represents the extension of antibody catalysis to a class of reactions for which no characterized enzyme exists.

5. ANTIBODIES THAT RECRUIT COFACTORS

A final strategy that has been applied to antibody catalysis involves the generation of antibodies that recruit cofactors. Many enzymes use cofactors to catalyse reactions, for example, cytochrome P450 (Fe-heme cofactor), α-ketoacid dehydrogenases (thiamine pyrophosphate cofactor), D-amino acid oxidase (flavin cofactor), and alanine racemase (pyridoxal phosphate cofactor). Strategies that allow incorporation of cofactors into antibody combining sites should expand the scope of antibody catalysis to redox reactions and energetically demanding hydrolytic reactions. The diversity of the antibody response should allow one to use both natural and unnatural cofactors.

In the past two years antibodies have been generated that bind both metal and redox active cofactors (Shokat *et al.* 1988; Iverson & Lerner 1989; Iverson *et al.* 1990). Most recently we have characterized an antibody that binds Fe(III)-mesoporphyrin IX and hydrogen peroxide and catalyses the oxidation of a number of substrates (Cochran & Schultz 1990*b*). Antibodies specific for N-methylporphyrins were subsequently found to form a stable complex with iron mesoporphyrin IX. The antibody Fe(III)–heme complex was found to catalyse the reductive breakdown of hydrogen peroxide in the presence of a variety of chromogenic electron donor substrates. The antibody-porphyrin complex remains active through at least 200 catalytic turnovers. Antibody catalysis could be completely inhibited by N-methylmesoporphyrin, and no catalysis was observed in the absence of the iron-mesoporphyrin cofactor. The catalysed peroxidation reaction followed Michaelis–Menten kinetics with respect to hydrogen peroxide reduction, with a K_M of 24 mM and k_{cat} of 394 min^{-1}. To generate a substrate binding site, in addition to the Fe(II)-heme and H_2O_2 sites, antibodies are now being generated to N-alkyl-porphyrins in which the alkyl group corresponds to the substrate of interest.

6. CONCLUSION

The number and diversity of reactions catalysed by antibodies continues to grow at a rapid rate. The specificity of antibody catalysed reactions is high and in some cases the rates of catalytic antibodies rival those of the corresponding enzyme catalysed reactions. The next few years will probably see an emphasis on increasing the catalytic efficiency of antibodies by generating antibodies that combine several catalytic mechanisms and by the use of genetic selections and screens. In addition, antibody catalysis is likely to be applied to reactions of biological, medical and chemical interest, for which naturally occuring enzymes do not exist.

We gratefully acknowledge the invaluable contributions of all our co-workers who are named in the citations. T.S.S. is supported by a Damon Runyon–Walter Winchell Cancer Research Fund Fellowship, DRG-1016, P.G.S. is a W.M.Keck Foundation Investigator. We acknowledge the National Institutes of Health for financial support.

REFERENCES

Bartlett, P. A. & Marlowe, C. K. 1983 Phosphonamidates as transition-state analogue inhibitors of thermolysin. *Biochemistry* **22**, 4618–4624.

Bennett, C. G. & Glaudemans, C. P. 1979 Contributions by ionic and steric features of ligands to their binding with phophorylcholine-specific immunoglobulin IgA H-8 as determined by fluorescence spectroscopy. *Biochemistry* **18**, 3337–3342.

Braisted, A. C. & Schultz, P. G. 1989 An antibody-catalyzed bimolecular Diels–Alder reaction. *J. Am. chem. Soc.* **112**, 7430–7431.

Crews, S., Griffin, J., Huang, H., Calame, K. & Hood, L. 1981 A single V_H gene segment encodes the immune response to phosphorylcholine: somatic mutation is correlated with the class of the antibody. *Cell* **25**, 59–66.

Cochran, A. G. & Schultz, P. G. 1990*a* Antibody-catalyzed porphyrin metallation. *Science, Wash.* **249**, 781–783.

Cochran, A. G. & Schultz, P. G. 1990*b* Peroxidase activity of an antibody-heme complex. *J. Am. chem. Soc.* **112**, 9414–9415.

Dailey, H. A. & Fleming, J. E. 1983 Bovine ferrochelatase, kinetic analysis of inhibition by N-methylprotoporphyrin, manganese, and heme. *J. biol. Chem.* **258**, 11453–11459.

Frick, L., Wolfenden, R., Smal, E. & Baker, D. C. 1986 Transition-state stabilization by adenosine deaminase: structural studies of inhibitory complex with deoxycoformycin. *Biochemistry* **25**, 1616–1621.

Goetze, A. M. & Richards, J. H. 1977 Magnetic resonance studies of the binding site interactions between phosphorylcholine and specific mouse myeloma immunoglobulin. *Biochemistry* **16**, 228–232.

Goetze, A. M. & Richards, J. H. 1978 Molecular studies of subspecificity differences among phophorylcholine-binding antibodies using ^{31}P nuclear magnetic resonance. *Biochemistry* **17**, 1733–1739.

Hilvert, D. & Nared, K. D. 1988 Stereospecific Claisen rearrangement catalyzed by an antibody. *J. Am. chem. Soc.* **110**, 5593–5594.

Hilvert, D., Hill, K. W., Nared, K. D. & Auditor, M. M. 1989 Antibody catalysis of a Diels–Alder reaction. *J. Am. chem. Soc.* **111**, 9261–9262.

Iverson, B. L. & Lerner, R. A. 1989 Sequence-specific peptide cleavage catalyzed by an antibody. *Science, Wash.* **243**, 1184–1188.

Iverson, B. L., Iverson, S. A., Roberts, V. A., Getzoff, E. D., Tainer, J. A., Benkovic, S. J. & Lerner, R. A. 1990 Metalloantibodies, *Science, Wash.* **249**, 659–662.

Jackson, D. Y., Jacobs, J. W., Sugasawara, R., Reich, S. H., Bartlett, P. A. & Schultz, P. G. 1988 An antibody catalyzed Claisen rearrangement. *J. Am. chem. Soc.* **110**, 4941–4942.

Jackson, D. Y., Prudent, J. R., Baldwin, E. P. & Schultz, P. G. 1991 A Mutagenesis study of a catalytic antibody. *Proc. natn Acad. Sci. U.S.A.* **88**, 58–63.

Jackson, D. Y. & Schultz, P. G. 1991 An antibody-catalyzed cis-trans isomerization. *J. Am. chem. Soc* (In the press.)

Janda, K. D., Lerner, R. A. & Tramantano, A. 1988 Antibody catalysis of bimolecular amide formation. *J. Am. chem. Soc.* **110**, 4835–4837.

Jencks, W. P. 1969 *Catalysis in chemistry and enzymology.* New York: McGraw-Hill Book Company.

Kohler, G. & Milstein, C. 1975 Continuous cultures of fused cells secreting antibody of predefined specificity. *Nature, Lond.* **256**, 495–497.

Lavallee, D. K. 1988 Porphyrin metallation reactions in biochemistry. *Molec. struct. Energ.* **9**, 279–314.

Napper, A. D., Benkovic, S. J., Tramantano, A. & Lerner, R. A. 1987 A stereospecific cyclization catalyzed by an antibody. *Science, Wash.* **237**, 1041–1043.

Page, M. I. & Jencks, W. P. 1971 Entropic contributions to rate accelerations in enzymic and intramolecular reactions and the chelate effect. *Proc. natn Acad. Sci. U.S.A.* **68**, 1678–1683.

Perlmutter, R., Crews, S., Douglas, R., Sorenson, G., Johnson, N., Nivera, N., Gearhart, P. & Hood, L. 1984 Diversity in phosphorylcholine-binding antibodies. *Adv. Immunol.* **35**, 1–37.

Pollack, S. J., Jacobs, J. W. & Schultz, P. G. 1986 Selective chemical catalysis by an antibody. *Science, Wash.* **234**, 1570–1573.

Pollack, S. J. & Schultz, P. G. 1987 Antibody catalysis by transitions state stablization. *Cold Spring Harb. Symp. quant. Biol.* **52**, 97–104.

Pollet, R., Edelhock, H., Rudikoff, S. & Potter, M. 1974 Changes in optical parameters of myeloma proteins with phosphorylcholine binding. *J. biol. Chem.* **249**, 5188–5194.

Rich, D. H. 1986 In *Proteinase inhibitors* (ed. A. J. Barrett & G. Salvesen). (179 pages.) Amsterdam: Elsevier.

Satow, Y., Cohen, G. H., Padlan, E. A. & Davies, D. R. 1986 Phosphocholine binding immunoglobulin Fab McPC603 an x-ray diffraction study at 2.7 Å. *J. molec. Biol.* **190**, 593–604.

Schultz, P. G. 1989 Catalytic antibodies. *Angew. Chem. Int. Ed. Engl.* **28**, 1283–1295.

Schultz, P. G., Lerner, R. A. & Benkovic, S. J. 1990 Catalytic antibodies. *Chem. Engng News* **68**, 26–40.

Shokat, K. M., Leumann, C. J., Sugasawara, R. & Schultz, P. G. 1988 An antibody-mediated redox reaction. *Angew. Chem. Int. Ed. Engl.* **27**, 1172–1174.

Shokat, K. M., Leumann, C. J., Sugasawara, R. & Schultz, P. G. 1989 A new strategy for the generation of catalytic antibodies. *Nature, Lond.* **338**, 269–271.

Shokat, K. M. & Schultz, P. G. 1990a Catalytic antibodies. *A. Rev. Immunol.* **8**, 335–363.

Shokat, K. M. & Schultz, P. G. 1990b The generation of antibody combining sites containing catalytic residues. *Ciba Foundation Symposium No. 159* (ed. J. Marsh). (In the press.)

Shokat, K. M., Ko, M. K., Scanlan, T. S., Kochersperger, L., Yonkovich, S., Thaisrivongs, S. & Schultz, P. G. 1990 Catalytic antibodies: a new class of transition state analogues used to elicit hydrolytic antibodies. *Angew. Chem. Int. Ed. Engl.* **291**, 1296–1303.

Tramantano, A., Janda, K. & Lerner, R. 1986 Catalytic antibodies. *Science, Wash.* **234**, 1566–1570.

Phil. Trans. R. Soc. Lond. B (1991)

The structure and function of protein modules

IAIN D. CAMPBELL AND MARTIN BARON

Department of Biochemistry, University of Oxford, South Parks Road, Oxford OX1 3QU, U.K.

SUMMARY

Analysis of protein sequences shows that many proteins in multicellular organisms have evolved by a process of exon shuffling, deletion and duplication. These exons often correspond to autonomously folding protein modules. Many extracellular enzymes have this modular structure; for example, serine proteases involved in blood-clotting, fibrinolysis and complement. The main role of these modules is to confer specificity by protein–protein interactions. Lack of structural information about such proteins has required a new strategy for studying the structure and function of protein modules. The strategy involves the production of individual modules by protein expression techniques, determination of their structure by high resolution nuclear magnetic resonance and definition of functional patches on the modules by site-directed mutagenesis and biological assays. The structures of the growth factor module, the fibronectin type 1 module and the complement module are briefly described. The possible functional roles of modules in various proteins, including the enzymes factor IX and tissue plasminogen activator, are discussed.

1. INTRODUCTION

There is, currently, an explosive growth in amino acid and nucleic acid sequence information. One of the striking features of these sequences is that, on analysis, certain patterns of amino acids, consensus sequences, appear over and over again. For example, a consensus sequence first recognized in epidermal growth factor (EGF) can be identified over 300 times in a wide variety of proteins. In many cases the domains or modules defined by the consensus sequences correspond to single exons and it is extremely likely that they have evolved and multiplied by a process of exon shuffling, insertion and duplication (Doolittle 1989; Patthy 1985, 1987). Modules are usually small, in the range 40–100 amino acids, and often fold autonomously (Baron *et al.* 1991).

The main biological role of protein modules seems to be for specific protein–protein interactions. They appear, for example, in cell surface molecules, such as receptors (Bazan 1990) and cell adhesion molecules (Bevilacqua *et al.* 1989; Springer 1990) and extra-cellular matrix proteins, such as fibronectin (Yamada 1989). They also appear in various serine proteases associated with blood clotting, fibrinolysis and complement (Neurath 1989); some members of the serum protease class of protein are shown in figure 1 and more information on the individual modules is given in the figure caption. The modular proteins found so far have been mainly extracellular although intracellular ones are also being found (Labeit *et al.* 1990).

In this paper, recent studies on the structure and function of some protein modules will be described with emphasis on proteins associated with blood clotting and fibrinolysis.

2. MODULAR PROTEINS AND THE STRUCTURES OF SOME MODULES

In general, proteins containing modules have been rather difficult to crystallize. This may be because they are usually glycosylated and possibly because of extensive flexibility in a protein consisting of a mosaic of modules (see figure 1). In addition, many of these proteins have only been available in rather small amounts and so were difficult to investigate by physical techniques. In recent years alternative approaches have become possible because of the relative ease with which one can produce proteins by the expression of heterologous genes using recombinant DNA. It is thus possible to produce fragments of modular proteins and determine their structure. This depends on the individual modules folding autonomously, but this often appears to occur. This approach has been used, for example, to investigate the structure of a pair of immunoglobulin modules from the cell-surface glycoprotein CD4 by X-ray diffraction methods (Wang *et al.* 1990; Ryu *et al.* 1990). As will be discussed here, recent advances in nuclear magnetic resonance (NMR) technology have also resulted in the determination of solution structures of a number of protein modules.

A strategy for studying protein modules has recently been outlined (Baron *et al.* 1991). The strategy can be summarized as follows. A module is identified in a sequence database; its gene is synthesized and expressed in a host cell such as yeast; the module is purified and its structure determined by NMR; this consensus structure can be used to model related modules; functional patches on the modules are recognized by sequence comparison, assay and site-specific mutagenesis; a model of the intact protein can be built up from low-resolution information from

Phil Trans. R. Soc. Lond. B (1991) **332**, 165–170
Printed in Great Britain

[59]

165

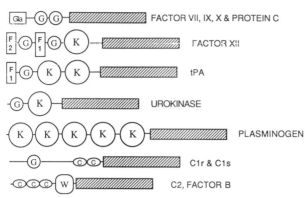

Figure 1. A diagram of the mosaic nature of some serine proteases found in serum. Factors VII, IX, X, XII and protein C are involved in blood clotting (Mann *et al.* 1988); tPA, urokinase and plasminogen are involved in fibrinolysis (Furie & Furie 1988); C1r, C1s, C2 and factor B are involved in complement (Law & Reid 1988).

The module Gla contains γ-carboxyglutamate and its structure has been determined by X-ray crystallography (Soriano-Garcia *et al.* 1989). The structure of the kringle module (K) has been determined both by diffraction (Soriano-Garcia *et al.* 1989, Harlos *et al.* 1987) and by NMR (Atkinson & Williams 1990). The structures of both the C module, also known as the short consensus repeat (SCR) and the complement control protein (CCP) module, and the type 1 module of fibronectin (F1) have been determined recently in this laboratory (Barlow *et al.* 1991; Baron *et al.* 1990). The G module is like epidermal growth factor (EGF) and its structure has been determined by NMR (see, for example, Campbell *et al.* 1990). The C7 module is found in various complement components such as C6 and C7 (Reid & Day 1990). The serine protease module (shaded box) is like trypsin (Neurath 1989) and the W module is found in the von Willebrand factor which is a multimeric plasma protein which associates with factor VIII (Mann *et al.* 1988; Reid & Day 1990).

microscopy plus information about how modules fit together, derived from studies of module pairs.

The technology associated with the determination of the structure of proteins in solution by using high resolution [¹H]-NMR is relatively recent (Wüthrich

1989; Bax 1989) but it is particularly convenient for this kind of study where relatively small polypeptides are involved. The current limit in protein size is around 150 amino acids, but this allows the determination of the structure of most modules and many module pairs. It is now possible to determine such structures in a relatively short time without the requirement of crystals or isomorphous heavy atom derivatives.

Figure 2 shows, schematically, structures of three modules derived in our laboratory. These were produced by using a yeast expression system that secreted the module into the cell medium (Baron *et al.* 1990). This was followed by analyses of high resolution NMR data and computation of families of structures consistent with the NMR restraints.

The G structure shown in figure 2 was derived from studies on human EGF (Cooke *et al.* 1987) and transforming growth factor alpha (TGF-α) (Tappin *et al.* 1989; Campbell *et al.* 1990). This module has three disulphide bridges and consists of a major double stranded antiparallel β-sheet with the three disulphides radiating up from one face of this sheet, attaching a rather ill-defined N-terminal strand and loop and a well-defined C-terminal region. (Lack of definition in NMR determined structures arises either from flexibility of the protein fragment or experimental difficulties; these two possibilities can often be distinguished by relaxation experiments.)

The dominant structural feature of the F1 module, which is about 40 residues long, is again antiparallel β-sheet (Baron *et al.* 1990). Two such sheets fold over a core of hydrophobic residues which are conserved in all members of the F1 module family. One of the two disulphide bonds joins these two sheet regions, the other stabilizes two adjacent strands of the main sheet.

The structure of the C module, which is about 60 amino acids long, has only been determined very recently (Barlow *et al.* 1991). This is again mainly formed from extended strands and β-sheet that fold around conserved hydrophobic residues.

It is worth emphasizing that, so far, evidence suggests that each member of a module family has the same

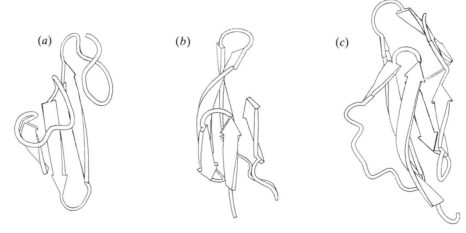

Figure 2. Schematic representations of the structures of three of the modules in figure 1. (*a*) shows the structure of a G module (Cooke *et al.* 1987); (*b*) shows the structure of an F1 module (Baron *et al.* 1990), and (*c*) shows the structure of a C module (Barlow *et al.* 1991). These structures were derived with NMR data collected from a solution containing modules produced by recombinant DNA techniques.

global fold but that different modules in a family can have different functional roles. The ability of proteins with a similar basic structure to bind a wide variety of molecules is well known for the immunoglobulin superfamily (Williams & Barclay 1988). It is becoming clear, however, that a similar multifunctional role is exhibited by other modules. Various very different functional patches have been identified on G modules; for example, a relatively large patch appears to be necessary on growth factors that bind to the EGF receptor (Campbell *et al.* 1990); an RGD sequence has been found to be involved in binding of cells to the basement membrane protein nidogen (Mann *et al.* 1989); a calcium-binding patch has been found in the N-terminal region of G modules from a wide variety of proteins including factor IX (Handford *et al.* 1990; Persson *et al.* 1989) and, in tissue plasminogen activator (tPA), yet another patch on a different part of the G structure has been implicated in the clearance of tPA from plasma by a liver receptor (Mark Edwards, British Biotechnology, private communication). The factor IX and tPA examples will be discussed in greater detail below.

It is of interest to consider how these different modules might fit together. The N- and C-termini of the C module are at different ends of the molecule and one might expect that this would facilitate a 'beads on a string' kind of assembly. The usual exon boundaries of the F1 and G modules, however, appear to leave uncompleted β-sheet structures at their C- and N-termini, respectively. It is thus tempting to speculate that these two modules might link up to complete a β-sheet structure (see discussion below on tPA and figure 4 *b*). In contrast to the three modules in figure 2, the N- and C-termini of the kringle module are close together in space. This would be expected to facilitate the formation of a coiled structure.

3. THE ROLE OF SOME MODULES IN BLOOD CLOTTING AND FIBRINOLYSIS

Vascular injury leads to the formation of an insoluble clot formed from fibrin and platelets (Mann *et al.* 1988; Furie & Furie 1988). This is brought about by a series of proteolytic enzyme reactions. This cascade of zymogen activations leads to the formation of thrombin and the subsequent cross-linking of a fibrin network. The cascade is tightly controlled by the presence of various regulatory proteins, e.g. factor VIII, and anticoagulant factors, e.g. protein C (Esmon 1989). In addition to clot formation, dissolution of clots is an important pathway and this, too, involves various zymogen activations leading to an attack on fibrin by plasmin (Furie & Furie 1988; Haber *et al.* 1989).

The role of modules will be briefly discussed in two examples here. One is the role of a G module and calcium ions in a complex associated with the formation of a blood clot, the other is the possible role of the F1-G module pair in tissue plasminogen activator, an enzyme which activates plasminogen.

(a) Modules in factor IX

Factor IX and factor X are similar serine protease

clotting factors (see figure 1). They are made up of a Gla module which contains modified glutamic acid residues, γ-carboxyglutamate, two G modules and a serine protease domain which is homologous to trypsin. The first G module usually contains either a β-hydroxyaspartate or a β-hydroxyasparagine. Factors IX and X form part of the coagulation cascade; the role of activated factor IX (IXa) being to activate factor X by cleaving its polypeptide chain. The chains remain attached after cleavage by means of a disulphide bridge.

Molecular defects in factor IX can lead to haemophilia B (Furie & Furie 1988; Lozier *et al.* 1990). Various experiments on fragments and different components have indicated that factors IXa and X form a complex with activated factor VIII (Mann *et al.* 1988). Factor VIII is a soluble plasma protein which binds with high affinity to certain phospholipid membranes and acts as a cofactor in the activation of factor X. Calcium ions are also required in this complex, as shown schematically in figure 3. Figure 3 is based on previously published schemes for the complex involving factor Xa, factor V and prothrombin (Harlos *et al.* 1987; Mann *et al.* 1988). This complex and the one shown in figure 3 are expected to be similar except that prothrombin contains kringle modules.

It is well known that the γ-carboxyglutamate residues in the Gla module bind calcium. Calcium ions appear not only to be associated with the Gla module, however, but also with the first G modules of factors IX and X. A consensus sequence D/N, D/N, D*/N*, Y/F (* denotes β-hydroxylation) has been recognized within this module (Rees *et al.* 1988), not only in factors IX and X but also in various other proteins including *notch* from *Drosophila* (Fehon *et al.* 1990). After obtaining the structure of a G module, Cooke *et al.* (1987) proposed that this sequence could form a calcium binding site, on one face of the G module. The putative calcium binding residues are shown in bold in figure 4 *a*, together with other 'structural' consensus residues of the G module shown in normal lettering. The importance of these residues is also implied by known point mutations in the first G module of factor

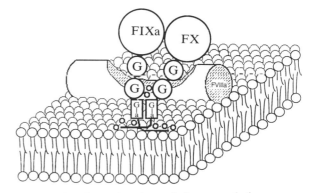

Figure 3. A schematic view of the coagulation enzyme complex involving factors VIIIa, IXa and X, calcium ions and phospholipid. Note the calcium ions (o) associated with the two G modules of factors IXa and X as well as those associated with the Gla modules and phospholipid.

Phil. Trans. R. Soc. Lond. B (1991)

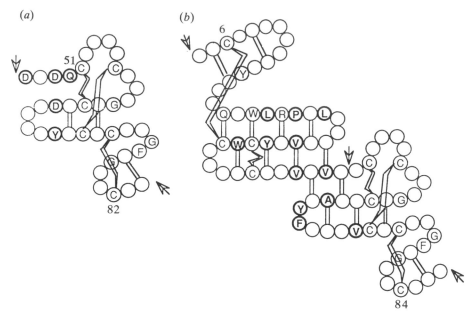

Figure 4. Structural representations of modules, showing the residue numbering in the intact proteins, the disulphide bonds, and possible H-bonding patterns. The consensus sequences are marked in normal lettering and bold letters indicate possible functional residues. The arrows show the exon boundaries of the F1 and G modules. (*a*) Corresponds to the first G module in factors IX and X, with the proposed calcium binding residues in bold. (*b*) Shows a possible structure of the F1-G module pair in tPA; a hydrophobic patch, formed by side-chains on one face of this model, is identified in bold.

IX which lead to haemophilia B (e.g. Lozier *et al.* 1990).

To explore in more detail the nature of this calcium binding site and its relationship to the complex in figure 3, we set out, in a collaboration with George Brownlee and his colleagues to define the role of the first G module of factor IX. This module was produced by recombinant methods and peptide synthesis. It was shown then that the NMR resonances of the tyrosine (Y69) in the consensus binding sequence were perturbed by calcium. This allowed an apparent K_d of 250 μM to be measured (Handford *et al.* 1990). In an independent study on the first G module produced from bovine factor X by proteolysis Persson *et al.* (1989) showed that this, too, bound calcium.

The next step was to obtain more precise information about residues involved in calcium binding by using site-directed mutagenesis. The effect of changed amino acids on calcium binding efficiency to the individual G modules was compared with the effect of such changes on the clotting efficiency of the intact protein (Handford *et al.* 1991). All the consensus binding residues have now been changed. A clear correlation between apparent K_d for calcium binding to the recombinant module and clotting efficiency of the intact protein was observed. For example, when the relatively conservative changes D47 → E and D64 → N were made, the calcium binding to the module was greatly reduced. These two mutations are known to cause haemophilia B and reduce clotting activity by a factor of about 100.

Another feature of investigations of module function is that there are large numbers of available module sequences, thus making sequence comparison a powerful tool (e.g. the prediction of a receptor binding patch

on EGF, Campbell *et al.* (1990)). A comparison of putative calcium binding sequences on other G modules (e.g. on *notch*, Fehon *et al.* (1990)) indicated that the position equivalent to D64 could be changed into N. In such modules, however, the residue equivalent to Q50 was usually E. This observation together with site-directed mutagenesis studies on recombinant G modules led to the realization that Q50 is important for the formation of the calcium binding site. This residue is therefore added to the consensus calcium binding patch shown in figure 4*a*.

These studies on this functional patch on a G module of factor IX indicate that the formation of the correct calcium binding site on the G module is very important in the complex shown in figure 3. Because the first G module on factor X has a very similar calcium binding site to the one on factor IX, it is not inconceivable that the two sites interact with each other in some way.

(*b*) *Modules in tissue plasminogen activator*

Tissue plasminogen activator (tPA) is involved in the removal of fibrin from the vascular system. The key enzyme in fibrinolysis is plasmin but the zymogen plasminogen must first be activated by a plasminogen activator. Of several such activators known, tPA appears to be the most specific for clot-associated plasminogen. The most likely mechanism for this specificity is a complex formed from plasminogen, fibrin and tPA (see, for example, Harris (1987); Haber *et al.* (1989); Higgins & Bennett (1990)).

Since tPA has considerable potential therapeutic and commercial value as a thrombolytic agent it has been extensively studied and many protein variants have been produced by recombinant methods. The main aims of these studies have been to improve fibrinolysis specificity, to increase the rate of thrombolysis and to increase the lifetime of the circulating protein (the *in vivo* half-life is only a few minutes).

tPA is composed of an F1 module, a G module, 2 K modules and a serine protease domain. There are also four potential N-glycosylation sites. More than 50 variant tPAs have been produced with module additions and deletions. The results have not always been clear cut and none of these variants has been shown to be superior to wild-type tPA in animal models (Higgins & Bennett 1990). It is generally agreed, however, that the F1 module and the second kringle are involved in binding to fibrin and that the G module is involved in clearance by a liver receptor.

To our knowledge no crystals of tPA suitable for X-ray studies have been produced, although much effort has gone into this by various groups. Before our studies, structures were available for the K module and the serine protease domain. Now that we also have knowledge of the consensus structures of F1 (Baron *et al.* 1990) and G (Campbell *et al.* 1990), structural information is available for all the components of the tPA mosaic. This does not, of course, give information about the overall shape of the molecule directly, since there are many possible ways the different components could fit together. It should, however, be possible to combine information about the individual module structures with information obtained from lower resolution methods, such as microscopy and scattering to refine models of the intact structure. Another way of obtaining this information might be to determine the structures of all the possible module pairs, e.g. F1-G, G-K and so on. In the case of tPA we have recently expressed an F1-G module pair for structural studies.

Until experimental information is available, one can only speculate about how any given pair of modules might fit together. However, as discussed in §2 above, the F1 and G modules appear to be constructed such that they have incomplete β-sheets at their C- and N-termini, respectively. In fact, the β-strands of the F1 module are rather like the G module in reverse (figure 4). It is therefore easy to visualize how the two modules might come together to complete these sheets as shown in figure 4*b*. A similar kind of rigid connection between two immunoglobulin modules has recently been reported in a study, by X-ray diffraction, of a recombinant fragment of CD4 (Wang *et al.* 1990). The kind of structure proposed in figure 4*b* would have one predominantly hydrophobic face and one predominantly hydrophilic face. One possibility is that the hydrophobic face could pack against the kringle modules, thus leading to a relatively compact structure for tPA. We expect to be able to produce a structure of the F1-G pair shortly and to combine this information with scattering data being collected by Colin Blake's group in Oxford to obtain a more complete picture of the solution structure of tPA.

4. CONCLUSIONS

The strategy outlined here, which involves the production of individual modules, the determination of

their structure by NMR and the definition of functional patches by mutagenesis and assay seems to be extremely powerful. The method has, so far, been applied successfully in this laboratory to various G modules, an F1 module and a C module. Numerous other modules and module pairs are currently under investigation and a rich harvest of information about modular proteins and their role in protein–protein interactions is expected soon.

This is a contribution from the Oxford Centre for Molecular Science which is supported by SERC and MRC. We also thank ICI Pharmaceuticals and British Biotechnology for their financial and technical support. The work owes much to many colleagues in Oxford, including Paul Barlow, Jonathan Boyd, Paul Driscoll, Tim Dudgeon, Tim Harvey, Uli Hommel and David Norman, who provided figure 2. We are particularly indebted to the George Brownlee group, especially Penny Handford and Mark Mayhew, with whom we collaborate on the work on FIX modules.

REFERENCES

Atkinson, A. & Williams, R. J. P. 1990 Solution structure of the kringle 4 domain from human plasminogen by ¹H NMR and distance geometry. *J. molec. Biol.* **212**, 541–552.

Barlow, P., Norman, D. G., Baron, M. & Campbell, I. D. 1991 The secondary structure of a complement control protein module by two-dimensional ¹H NMR. *Biochemistry* **30**, 997–1004.

Baron, M., Norman, D. & Campbell, I. D. 1991 Protein modules *TIBS.* **16**, 13 17.

Baron, M., Norman, D. G., Willis, A. & Campbell, I. D. 1990 Structure of the fibronectin type 1 module. *Nature, Lond.* **345**, 642–646.

Bax, A. 1989 2D NMR and protein structure. *A. Rev. Biochem.* **58**, 223–256.

Bazan, J. F. 1990 Structural design and molecular evolution of a cytokine receptor superfamily. *Proc. nat. Acad. Sci., U.S.A.* **87**, 6934–6938.

Bevilacqua, M. P., Stengelin, S., Gimbrone, M. A. & Seed, B. 1989 Endothelial leukocyte adhesion molecule 1: an inducible receptor for neutrophils related to complement regulatory proteins and lectins. *Science, Wash.* **243**, 1160–1165.

Campbell, I. D., Cooke, R. M., Baron, M., Harvey, T. S. & Tappin, M. J. 1989 The solution structures of EGF and TGF-α. *Prog. grow. Fact. Res.* **1**, 13–22.

Campbell, I. D., Baron, M., Cooke, R. M., Dudgeon, T. J., Fallon, A., Harvey, T. S. & Tappin, M. J. 1990 Structure function relationships in EGF and TGF-α. *Biochem. Pharmac.* **40**, 35–40.

Cooke, R. M., Wilkinson, A. J., Baron, M., Pastore, A., Tappin, M. J., Campbell, I. D., Gregory, H. & Sheard, B. 1987 The solution structure of human epidermal growth factor. *Nature, Lond.* **327**, 339–341.

Doolittle, R. F. 1989 Similar amino-acids revisited. *TIBS* **14**, 244–245.

Dudgeon, T. J., Baron, M., Cooke, R. M., Campbell, I. D., Edwards, R. M. & Fallon, A. 1990 Structure and function of hEGF: receptor binding and NMR. *FEBS Lett.* **261**, 392–396.

Esmon, C. T. 1989 The roles of protein C and thrombomodulin in the regulation of blood coagulation. *J. biol. Chem.* **264**, 4743–4746.

Fehon, R. G., Kooh, P. J., Rebay, I., Regan, C. L., Xu, T., Muskavitch, M. A. T. & Artavanis-Tsakonas, S. 1990 Molecular interactions between the protein products of the neutrogenic loci *notch* and *delta*, two EGF-homologous genes in *Drosophila. Cell* **61**, 523–534.

Furie, B. & Furie, B. C. 1988 The molecular basis of blood coagulation. *Cell* **53**, 505–517.

Haber, E., Quertermous, T., Matsueda, G. R. & Runge, M. S. 1989 Innovative approaches to plasminogen activator therapy. *Science, Wash.* **243**, 51–56.

Handford, P. A., Baron, M., Mayhew, M., Willis, A., Beesley, T., Brownlee, G. G. & Campbell, I. D. 1990 The first EGF-like domain of human factor IX has a high affinity Ca^{++} binding site. *EMBO J.* **9**, 475–480.

Handford, P. A., Mayhew, M., Baron, M., Winship, P. R., Campbell, I. D. & Brownlee, G. G. *Nature, Lond.* (In the press.)

Harris, T. J. R. 1987 Second generation plasminogen activators. *Protein Engineering* **1**, 449–458.

Harlos, K., Holland, S. K., Boys, C. W. G., Burgess, A. I., Esnouf, M. P. & Blake, C. C. F. 1987 Vitamin K-dependent blood coagulation proteins form heterodimers. *Nature, Lond.* **330**, 82–84.

Higgins, D. L. & Bennett, W. F. 1990 Tissue plasminogen activator: the biochemistry and pharmacology of variants produced by mutagenesis. *A. Rev. Pharmac. Toxicol.* **30**, 91–121.

Labeit, S., Barlow, D. P., Gautel, M., Gibson, T., Holt, J., Hsieh, Francke, U., Leonard, K., Wardale, J., Whiting, A. & Trinick, J. 1990 A regular pattern of two types of 100-residue motif in the sequence of titin. *Nature, Lond.* **345**, 273–276.

Law, S. K. A. & Reid, K. B. M. 1988 *Complement* Oxford: IRL Press.

Lozier, J. N., Monroe, D. M., Stanfield-Oakley, S., Lin, S.-W., Smith, K. J., Roberts, H. R. & High, K. A. 1990 Factor IX New London: substitution of proline for glutamine at position 50 causes several hemophilia B. *Blood* **75**, 1097–1104.

Mann, K., Deutzmann, Aumailley, M., Timpl, R., Raimondi, L., Yamada, Y., Pan, T., Conway, D. & Chu, M.-L. 1989 Amino-acid sequence of mouse nidogen, a multidomain basement membrane protein with binding activity for laminin, collagen IV and cells. *EMBO J.* **8**, 65–72.

Mann, K. G., Jenny, R. J. & Krishnaswamy, S. 1988 Cofactor proteins in the assembly and expression of blood clotting enzyme complexes. *A. Rev. Biochem.* **57**, 915–956.

Neurath, H. 1989 Proteolytic processing and physiological regulation. *TIBS* **14**, 268–271.

Patthy, L. 1985 Evolution of the proteases of the blood clotting and fibrinolysis by assembly from modules. *Cell* **41**, 657–663.

Patthy, L. 1987 Intron-dependent evolution: preferred types of exons and introns. *FEBS Lett.* **214**, 1–7.

Persson, E., Selander, M., Linse, S., Drakenberg, T., Ohlin, A. & Stenflo, J. 1989 Calcium binding to the isolated β-hydroxyaspartic acid-containing epidermal growth factor-like domain of bovine factor X. *J. biol. Chem.* **264**, 16897–16904.

Rees, D. J. G., Jones, I. M., Handford, P. A., Walter, S. J., Esnouf, M. P., Smith, K. J. & Brownlee, G. G. 1988 The role of β-hydroxyaspartate and adjacent carboxylate residues in the first EGF domain of human factor IX. *EMBO J.* **7**, 2053–2061.

Reid, K. B. M. and Day, A. J. 1989 Structure function relationships of the complement system. *Immun. Today* **10**, 177–180.

Ryu, S. E., Kwong, P. D., Truneh, A., Porter, T. G., Arthos, J., Rosenberg, M., Dai, X., Xuong, N., Axel, R., Sweet, R. W. & Hendrickson, W. A. 1990 Crystal structure of HIV-binding recombinant fragment of human CD4. *Nature, Lond.* **348**, 419–426.

Soriano-Garcia, M., Park, C. H., Tulinsky, A., Ravichandran, K. G. & Skrzypczak-Jankun, E. 1989 Structure of Ca^{++} prothrombin fragment 1 including the conformation of the gla domain. *Biochemistry* **28**, 6805–6810.

Springer, T. A. 1990 Adhesion molecules of the immune system. *Nature, Lond.* **346**, 425–434.

Tappin, M. J., Cooke, R. M., Fitton, J. & Campbell, I. D. 1989 A high resolution ^{1}H NMR study of hTGF-α: structure and pH dependent conformational interconversion. *Eur. J. Biochem* **179**, 629–637.

Wang, J., Yan, Y., Garrett, T. P. J., Liu, J., Rodgers, D. W., Garlick, R. L., Tarr, G. E., Husain, Y., Reinherz, E. L. & Harrison, S. C. 1990 Atomic structure of a fragment of human CD4 containing two immunoglobulin-like domains. *Nature, Lond.* **348**, 411–418.

Williams, A. F. & Barclay, A. N. 1988 The immunoglobulin superfamily – domains for cell surface recognition. *A. Rev. Immunol.* **6**, 381–405.

Wüthrich, K. 1989 Protein structure determination in solution by NMR. *Science, Wash.* **243**, 45–50.

Yamada, K. M. 1989 Fibronectins: structure functions and receptors. *Curr. Opin. Cell Biol.* **1**, 956–963.

Pathway and stability of protein folding

ALAN R. FERSHT, MARK BYCROFT, AMNON HOROVITZ,
JAMES T. KELLIS Jr, ANDREAS MATOUSCHEK AND LUIS SERRANO

MRC Unit for Protein Function and Design, Department of Chemistry, University of Cambridge, Lensfield Road, Cambridge CB2 1EW, U.K.

SUMMARY

We describe an experimental approach to the problem of protein folding and stability which measures interaction energies and maps structures of intermediates and transition states during the folding pathway. The strategy is based on two steps. First, protein engineering is used to remove interactions that stabilize defined positions in barnase, the RNAse from *Bacillus amyloliquefaciens*. The consequent changes in stability are measured from the changes in free energy of unfolding of the protien. Second, each mutation is used as a probe of the structure around the wild-type side chain during the folding process. Kinetic measurements are made on the folding and unfolding of wild-type and mutant proteins. The kinetic and thermodynamic data are combined and analysed to show the role of individual side chains in the stabilization of the folded, transition and intermediate states of the protein. The protein engineering experiments are corroborated by nuclear magnetic resonance studies of hydrogen exchange during the folding process. Folding is a multiphasic process in which α-helices and β-sheet are formed relatively early. Formation of the hydrophobic core by docking helix and sheet is (partly) rate determining. The final steps involve the forming of loops and the capping of the N-termini of helices.

1. INTRODUCTION

Molecular recognition in biology is principally controlled by non-covalent bonds: van der Waal's. electrostatic, hydrogen bonding and the hydrophobic effect. These interactions are responsible for information transfer from DNA to RNA to proteins, the assembly of macromolecules, the binding of ligands and the three-dimensional structure of proteins. The essence of enzymic catalysis is the use of binding energy and complementarity of enzyme–substrate interactions (Fersht 1985). It is now possible to study non-covalent interactions within proteins directly by using protein engineering (Winter *et al.* 1982). We are now applying protein engineering methods to the problem of protein folding: the prediction of the three-dimensional tertiary structure of a protein from its linear sequence of amino acids. Protein folding, like any chemical process, has two components, kinetic and thermodynamic: the pathway of folding and the stability of the folded state. It is not yet possible to solve either of these components by computational methods *de novo*. As far as kinetics is concerned, too many conformations occur between the denatured protein and the folded structure to be searched at random. It is not even possible today to calculate whether a known folded structure is stable with respect to its denatured state. Experimental data are necessary for both the thermodynamic and kinetic processes.

We have developed a strategy using protein engineering to provide necessary experimental data on both pathway and stability (Matouschek *et al.* 1989,

1990). First, we use site-directed mutagenesis to remove interactions that stabilize parts of proteins and then measure the changes in stability. This provides empirical thermodynamic data that can be used in redesigning enzymes and provides a data base for testing theoretical procedures. Second, we perform kinetic measurements on folding and unfolding of the mutated enzymes to measure the changes in activation energies and equilibrium energy levels. The relationship between changes in activation and equilibrium energies may be used to map the structures of transition states and intermediates (Matouschek *et al.* 1989, 1990; Matouschek & Fersht 1991).

Our protein of choice as a paradigm for protein folding studies is barnase, an extracellular ribonuclease from *Bacillus amyloliquefaciens*. It is a small monomeric enzyme of 110 residues, with a relative molecular mass of M_r of 12382. It is composed of a single domain of about the size expected for a folding unit of a large protein, and has α and β secondary structure. Barnase undergoes reversible solvent- and thermally induced denaturation, closely approximating to a two-state transition. The crystal structure of the protein has been solved at high resolution (Mauguen *et al.* 1982) and its solution structure has also been elucidated by high-field proton nuclear magnetic resonance (NMR) spectroscopy (M. Bycroft, unpublished data). Importantly for kinetic experiments, there are minimal effects from the rates of proline isomerization. The kinetics of folding of proteins is often complicated by the slow and frequently rate-determining isomerization of proline residues (Schmid *et al.* 1986). It is expected that some

Phil. Trans. R. Soc. Lond. B (1991) **332**, 171–176
Printed in Great Britain

[65]

171

12-2

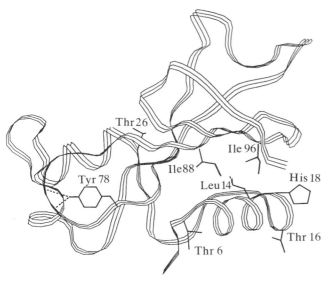

Figure 1. Sketch of barnase, showing the initial residues mutated in this study.

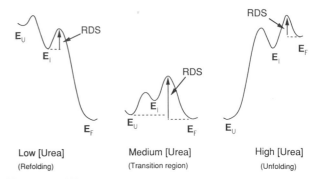

Figure 2. Minimal free energy profiles that describe the unfolding and refolding of barnase at low (< 2 M), medium (4–5 M) and high (> 6 M) [urea].

5–20 % of each of the prolines in denatured protein should exist in the *cis* conformation and that the conformations should interconvert with a half-life of tens of seconds (Brandts *et al.* 1975). Barnase has just three proline residues and they are all *trans* in the native structure. Consequently, the major phase in the refolding of barnase is the fraction ($\approx 75\%$) of unfolded protein that has all its prolines in the *trans* conformation in solution. The gene for this enzyme has been cloned and can be expressed in *Escherichia coli* (Paddon & Hartley 1987). The tertiary structure of barnase comprises a C-terminal five-stranded anti-parallel β-sheet (residues 50–55, 70–75, 85–91, 94–101 and 106–108) with two N-terminal α-helices, the major one (residues 6–18) packed against its face and the other (26–34) against its edge. The hydrophobic core of barnase is formed from the non-polar side chains of the major α-helix interdigitating with those of the β-sheet (figure 1).

2. EXPERIMENTAL APPROACH

Protein denaturation is monitored by the large decrease in the fluorescence of the tryptophans. Our main method of inducing denaturation is by the addition of urea. The data are extrapolated to 0 M [urea] by standard procedures (Pace 1986). We find the results thus obtained are independent of whether denaturation is urea-, guanidinium- or temperature-induced (Kellis *et al.* 1989). Kinetics of denaturation and renaturation are obtained by stopped-flow fluorescence and [urea]-jumps (Matouschek *et al.* 1989, 1990).

3. FOLDING INTERMEDIATE

An intermediate, or series of intermediates, was detected in the refolding of barnase (Matouschek *et al.* 1990). The intermediate does not appreciably accumulate at medium (4–5 M) to high concentrations of

urea, but does do so at low concentrations (figure 2). There is growing evidence that the refolding of proteins *in vitro* may proceed, in general, *via* transient intermediates (Udgaonkar & Baldwin 1988; Roder *et al.* 1988). Earlier work has shown the existence of a state, termed the molten globule, which is stable under mildly denaturing conditions (acid pH, moderate concentrations of denaturants, or high temperatures (Ptitsyn *et al.* 1990)). An outstanding question is what is the structure of these intermediates, and do they have the common features for the molten globule. We can analyse the structure of the intermediate and also the transition state for its formation from the folded state (the 'unfolding' transition state) by protein engineering methods (Matouschek *et al.* 1989, 1990; Matouschek & Fersht 1991).

4. DIFFERENCE ENERGY DIAGRAMS

It was shown how to analyse the role of binding energy of groups at the active site of the enzymes by determining the free energy profiles of wild type and mutants throughout the reaction, and measuring the differences between the two (Wells & Fersht 1986; Fersht *et al.* 1987). The 'difference energy' diagrams readily give a qualitative, and even quantitative (Fersht 1988), picture of the contribution of each side chain to binding and, hence catalysis, throughout the reaction. An identical approach is now used to analyse protein folding. First, suitable mutations must be chosen. Ideally, the target side chain is involved in just one interaction with one other side chain, preferably a weak interaction. The moiety that makes that interaction is then deleted by mutagenesis. Suitable changes are Ile → Val, Thr → Ser and Tyr → Phe. It is important not to remove a residue that solvates a buried charge as the charged residue in the mutant may cause serious structural changes in becoming solvated. If these mutations are non-disruptive, then the changes in energy on mutation reflect the interaction of the target side chain. Next, the free energy profiles of the folding pathway wild-type must be constructed. The kinetics of folding and unfolding fit that expected for a simple two-step reaction but this is just part of a more general scheme which can have multiple intermediates. The intermediate we see is the slowest formed, or a mixture of such intermediates.

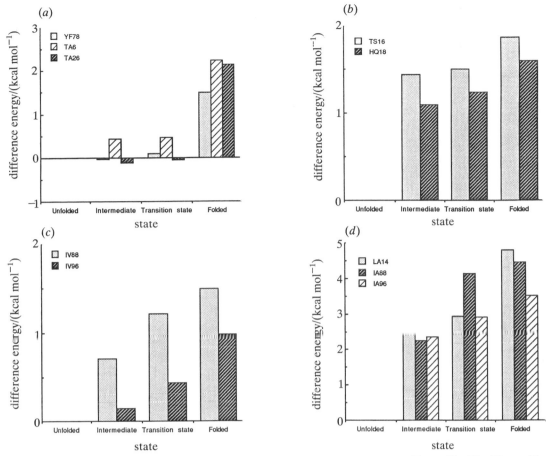

Figure 3. Difference energy diagrams for the folding of the mutants: (*a*) Tyr → Phe 78, Thr → Ala 6 and Thr → Ala 26; (*b*) Thr → Ser 16 and His → Gln 18; (*c*) Ile → Val 88 and Ile → Val 96; Leu → Ala 14, Ile → Ala 88 and Ile → Ala 96.

5. INITIAL DESCRIPTION OF INTERMEDIATE AND UNFOLDING TRANSITION STATE FROM PROTEIN ENGINEERING METHODS

Difference energy diagrams for a variety of mutants, each chosen to probe specific points of the structure, are plotted in figure 3. We sometimes describe the changes in difference energies in terms of a function ϕ. For example, if the change in free energy of unfolding on mutagenesis is $\Delta\Delta G_U$, and the change in activation energy on mutation for the rate constant of unfolding is $\Delta\Delta G_U^{\ddagger}$, then $\phi_U = \Delta\Delta G_U^{\ddagger}/\Delta\Delta G_U$. Quantitative interpretation is complicated because the unfolded state is the reference structure for each mutant. The difference energies are not, therefore, the true differences in non-covalent energy between the wild-type and mutant enzymes because the measured energies for each mutant differ from the true by a term which is the difference in free energy of the unfolded wild-type protein and the specific mutant (Matouschek *et al.* 1990; Matouschek & Fersht 1991). This does not, however, affect the interpretation of two extreme, but very common, cases: (i) the target side chain is as exposed in the transition state or intermediate as it is in the unfolded protein, the difference energy = 0 = ϕ; (ii) the target side chain makes the same interaction in the transition state or intermediate as it does in the folded protein, the difference energy is thus the same as

in the folded state, and $\phi = 1$. The interpretation of fractional values of ϕ is affected by the differences in energy of the unfolded states but we have shown where this may be ignored or where the changes can be analysed more quantitatively (Matouschek *et al.* 1989, 1990; Matouschek & Fersht 1991). The roles of the different side chains in stabilizing the different states are seen immediately on examining the difference energy diagrams (figure 3).

(*a*) *Formation of the N-termini of the helices and a major loop is a very late step in folding*

Tyr 78 stabilizes a major loop by the —OH group forming hydrogen bonds with the > NH and > C = O of Gly 81 in the folded state. It is seen on mutation of Tyr → Phe 78 that all the stabilization energy of the hydrogen bonds is lost in both the transition state for unfolding and in the intermediate. The —OH of Thr 26 acts as the N-cap of the second helix, forming hydrogen bonds to the > NHs of residues 27–29 (Serrano & Fersht 1989). On mutating Thr → Ala 26, it is seen that all the energy of the N-cap is lost in the transition state for unfolding, and the same is seen to happen in the intermediate. Thr 6 forms the N-cap of the first helix. On mutating Thr → Ala 6, 80% of the difference energy is lost in the transition state for unfolding and the same energy is lost in the intermediate as in the transition state.

Phil. Trans. R. Soc. Lond. B (1991)

(b) *Formation of the C-terminus of the major α-helix is an early event in folding*

The charge on the protonated form of His 18 makes a coulombic interaction with the dipole of the α-helix from residues 6–18 as well as there being a specific hydrogen bond between an imidazole > NH and the > CO of Gln 15. The interaction energy is thus a probe of the integrity of the C-terminus of the helix. Most of this interaction energy is maintained in the transition state and in the intermediate. The γ-methyl of Thr 16 makes a very strong hydrophobic interaction with the aromatic ring of Tyr 17. This is maintained in the transition state for unfolding and in the intermediate. Both probes give consistent results.

(c) *Formation of the hydrophobic core is (partly) rate determining*

Mutation of Ile → Val 96 and Ile → Val 88 probes the hydrophobic core. Some of the energy of the hydrophobic core is lost in the transition state for unfolding, possibly 10–30% depending on the particular location. Even more energy is lost in the intermediate. The energy changes in the intermediate are midway between those of the unfolded and folded states. In this situation, the changes in difference energies do reflect the true energy changes (Matouschek *et al.* 1989). Consolidation of the hydrophobic core is part of the rate determining process.

6. EXTENSION OF PROTEIN ENGINEERING APPROACH: COSMIC ANALYSIS

There are two limitations in the approach described so far. The first is that it works best for mutation of just simple interactions. Adventitious structural changes may occur when many interactions are broken on mutation and cause a reorganization energy term in the equations. The second is that we assume that the retention of an interaction energy implies retention of the specific interaction. Although this will be true in most cases, it could be that different interactions of similar energy to those in the folded state are taken up in other states. We have now extended the range and reliability of the protein engineering procedure by using the cosmic (Combination of Sequential Mutant Interaction Cycles) technique in which a series of double-mutant cycles is constructed. In each cycle, the side chains of two amino acid residues that interact in the folded state are mutated separately and together. The technique of double-mutant cycles was introduced to detect interactions between groups at the active site of the tyrosyl-tRNA synthetase (Carter *et al.* 1984; Lowe *et al.* 1985). A formal analysis has been given for their rigorous application to energy changes on protein folding (Serrano *et al.* 1990). Two or more (Horovitz & Fersht 1990), residues, X and Y, that interact are identified from examination of the structure of protein. The two residues are mutated separately and then together to give a cycle comprising E-XY, EX, EY and E. In ideal circumstances, where the mutations cause no rearrangement in the protein structure, subtraction

of the free energy change for E-X → E($\Delta G_{E-X \to E}$) from the free energy change for E-XY → EY($\Delta G_{E-XY \to EY}$) causes the interaction energies of X and Y with the rest of the protein to cancel out. An interaction energy, $\Delta G_{int} = \Delta G_{E-XY \to EY} - \Delta G_{E-X \to E}$, is obtained which is equal to

$$\Delta G_{int} = G_{X \cdots Y} - \Delta G_{X \cdots w} - \Delta G_{Y \cdots w}, \qquad (1)$$

where $G_{X \cdots Y}$ is the interaction energy between X and Y, $\Delta G_{X \cdots w}$ is the increase in solvation energy of X on removal of Y, and $G_{Y \cdots w}$ is the increase in solvation energy of Y on removal of X (Serrano *et al.* 1990). Equation (1) still holds if there is disruption of the structure of the enzyme on mutation of, say, X in E-XY but the same disruption occurs on the mutation of X in E-X. Structural artefacts on mutation thus tend to cancel out, reducing their importance to second-order effects.

A quantity, ϕ_{int}, may be defined that is equal to $\Delta\Delta G_{int(X)}/\Delta\Delta G_{int(F)}$, where $\Delta\Delta G_{int(F)}$ is the value of $\Delta\Delta G_{int}$ in the folded state and $\Delta\Delta G_{int(X)}$ is that in any other state on the folding pathway. $\phi_{int} = 1$ means that the inteaction energy is fully maintained in state X whereas $\phi_{int} = 0$ indicates complete loss. Importantly, fractional values of ϕ_{int} may be simply interpretable. Where the interaction arises from a direct contact, such as a van der Waal's interaction, the two residues have to be within close contact for a significant energetic contribution. Thus appreciable values of $\Delta\Delta G_{int}$ from the double-mutant cycles show unambiguously that the residues continue to interact in the other state, X. It is possible, however, that the mode of interaction changes.

This approach has been applied to the major helix of barnase (Horovitz *et al.* 1991). The γ-methyl group of Thr 16 interacts with the face of the aromatic ring of Tyr 17 which, in turn, makes an aromatic–aromatic interaction with the side chain of Tyr 13 (Serrano *et al.* 1991). The side chain of Asp 12 is linked to that of Asp 8 via a salt bridge with the guanidinium group of the side chain of Arg 110, the C-terminal residue as shown in figure 2 (Horovitz *et al.* 1990). Tyr 17 makes a large number of interactions within the helix with Tyr 13 and Thr 16. Mutation to Ala 17 is, accordingly, very radical. Tyr 13, as well as its interaction with Tyr 17, makes extensive interactions with the side chain of Pro 21, the main chain atoms of Lys 19, Leu 20 and Pro 21. Mutation of Tyr → Ala 13 is thus a radical change that removes many interactions, including those with residues outside the helix. Mutation of Arg 110 breaks bonds to both Asp 8 and Asp 12. Mutation of Asp 8 affects the energetics of the Asp 12–Arg 110 interaction whereas mutation of Asp 12 affects the energetics of the Asp 8–Arg 110 interaction (Horovitz *et al.* 1990). It is seen in figure 4, however, that the results of the double-mutant cycles are quite clearcut. The value of the energy of the Tyr–Tyr interaction is hardly changed on going from the folded to the transition to the intermediate. This shows that the centre of the helix is already fully formed in the folding intermediate on the folding pathway. The Asp 12–Arg 110 interaction is weakened somewhat during unfolding, and the Asp 8–Arg 110

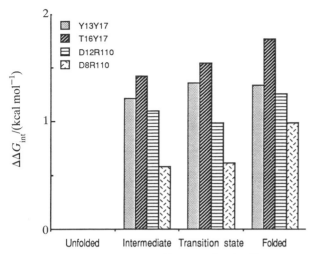

Figure 4. Difference energy diagrams for double mutant cycles (ϕ_{int}).

somewhat more so. This is further evidence that the N-terminus of the helix unwinds first during unfolding or forms late in folding.

7. DETECTION AND CHARACTERIZATION OF THE INTERMEDIATE BY NMR

Recently, NMR experiments has been used to detect folding intermediates. NMR can detect backbone $> NH$ groups that undergo H/D exchange slowly with solvent because they are hydrogen-bonded within α-helices, β-sheets or other structural elements. Rapid quenching experiments on H/D exchange during refolding can detect secondary structure that is formed faster than is the final folded structure. For example, regions of α-helix and β-sheet are formed rapidly in the refolding of RNase A (Udgaonkar & Baldwin 1988) and in cytochrome *c* (Roder *et al.* 1988). More than 99 % of the proton resonances in the ^{1}H-NMR spectrum of barnase have been assigned to their amino acid residues (Bycroft *et al.* 1990) and so we have used the H/D exchange procedure to detect a rapidly formed intermediate during refolding. Barnase was denatured and all exchangeable $> NH$ groups deuterated ($> 90\%$ exchange) by incubating in 6.5 M (deuterated) urea and 99.8 % D_2O at pD 6.3. The denatured and deuterated protein was allowed to refold by diluting into D_2O. Samples were taken during the refolding process and exposed to a labelling pulse of H_2O buffered at pH 8.5 where there is fast exchange of unprotected $> ND$ deuterons. After sufficient time for complete refolding, the pH was lowered to 3.5 where exchange is very slow, and the fraction of H/D exchange measured by 2-D NMR. If folding is a simple one-step process, then the $> ND$ deuterons at positions which undergo slow exchange in the folded protein should become protected from exchange concurrently with the overall process of refolding, which has a half life of 140 ms under these conditions in D_2O and 1.3 M urea. It is found, however, that several of these deuterons become protected with a half-life of about 12–30 ms. The rapidly protected positions are in the secondary structure of the two α-helices and the

five strands of β-sheet. Three of the positions, the $> ND(H)$s of I25, N77 and S50, are protected according to a timecourse that is essentially identical to that of the overall refolding process. Significantly, I25, N77 and S50 make tertiary hydrogen bonds that are not within regular secondary structure but are a consequence of the overall fold of the molecule. This shows conclusively folding is multiphasic process in which an intermediate is rapidly formed. The data are consistent with the intermediate possessing some of the α and β secondary structure that is present in the final folded state.

8. DESCRIPTION OF INTERMEDIATE AND IMPLICATIONS

The NMR method gives information on just the secondary structure whereas protein engineering gives evidence, both quantitative and qualitative, on the interaction of side chains and indirect evidence about secondary structure. As well as the two procedures complementing each other, there is overlapping information that can be crosschecked. In every case, we find that there is complete agreement between NMR and protein engineering results. Both sets of experiments show that the transition state for unfolding and the folding intermediate have considerable secondary structure. The strands in the β-sheet are mainly intact, but loops are unfolded. The C-termini of the helices are relatively intact but the N-termini are unfolded. The hydrophobic core is weakened in the transition state for unfolding and even more so in the intermediate. One proposal for the prediction of the tertiary structure of proteins is to predict the repeating secondary structure, such as α-helices and β-sheets, which is less demanding than predicting tertiary structure, and then dock the helices and sheets. Our results suggest that docking of preformed elements is part of the rate determining process in folding and so this provides encouragement for that theoretical approach.

REFERENCES

Brandts, J. F., Halvorson, H. R. & Brennan, M. 1975 *Biochemistry* **14**, 4953–4963.

Carter, P. J., Winter, G., Wilkinson, A. J. & Fersht, A. R. 1984 *Cell, Camb., Mass.* **38**, 835–840.

Bycroft, M., Matouschek, A., Kellis, J. T. Jr, Serrano, L. & Fersht, A. R. 1990 *Nature, Lond.* **346**, 488–490.

Fersht, A. R. 1985 *Enzyme structure and mechanism* (2nd edn). New York: W. H. Freeman.

Fersht, A. R. 1988 *Biochemistry* **27**, 1577–1580.

Fersht, A. R., Leatherbarrow, R. J. & Wells, T. N. C. 1987 *Biochemistry* **26**, 6030–6038.

Horovitz, A. & Fersht, A. R. 1990 *J. molec. Biol.* **214**, 613–617.

Horovitz, A., Serrano, L., Avron, B., Bycroft, M. & Fersht, A. R. 1990 *J. molec. Biol.* **216**, 1031–1044.

Kellis, J. T. Jr, Nyberg, K. & Fersht, A. R. 1989 *Biochemistry* **28**, 4914–4922.

Lowe, D. M., Fersht, A. R., Wilkinson, A. J., Carter, P. & Winter, G. 1985 *Biochemistry* **24**, 5106–5109.

Matouschek, A. & Fersht, A. R. 1991 *Methods Enzymol.* (In the press.)

Matouschek, A., Kellis, J. T. Jr, Serrano, L. & Fersht, A. R. 1989 *Nature, Lond.* **340**, 122–126.

Matouschek, A., Kellis, J. T. Jr, Serrano, L., Bycroft, M. & Fersht, A. R. 1990 *Nature, Lond.* **346**, 440–445.

Mauguen, Y., Hartley, R. W., Dodson, E. J., Dodson, G. G., Bricogne, G., Chothia, C. & Jack, A. 1982 *Nature, Lond.* **297**, 162–164.

Pace, C. N. 1986 *Methods Enzymol.* **131**, 266–279.

Paddon, C. J. & Hartley, R. W. 1987 *Gene* **53**, 11–19.

Ptitsyn, O. B., Pain, R. H., Semisotnov, G. V., Zerovnik, E. & Razgulyaev, O. I. 1990 *FEBS Lett.* **262**, 20–24.

Roder, H., Elöve, G. A. & Englander, S. W. 1988 *Nature, Lond.* **335**, 700–704.

Schmid, F. X., Grafl, R., Wrba, A. & Beintema, J. J. 1986 *Proc. natn. Acad. Sci. U.S.A.* **83**, 872–876.

Serrano, L., Bycroft, M. & Fersht, A. R. 1991 *J. molec. Biol.* (In the press.)

Serrano, L. & Fersht, A. R. 1989 *Nature, Lond.* **342**, 296–299.

Serrano, L., Horovitz, A., Avron, B., Bycroft, M. & Fersht, A. R. 1990 *Biochemistry* **29**, 9343–9352.

Udgaonkar, V. & Baldwin, R. L. 1988 *Nature, Lond.* **335**, 694–699.

Wells, T. N. C. & Fersht, A. R. 1986 *Biochemistry* **25**, 1881–1886.

Winter, G., Fersht, A. R., Wilkinson, A. J., Zoller, M. & Smith, M. 1982 *Nature, Lond.* **299**, 756–758.

Design and synthesis of new enzymes based on the lactate dehydrogenase framework

C. R. DUNN[1], H. M. WILKS[1], D. J. HALSALL[1], T. ATKINSON[2], A. R. CLARKE[1], H. MUIRHEAD[1] AND J. J. HOLBROOK[1]

[1] *University of Bristol Molecular Recognition Centre, School of Medical Sciences, University Walk, Bristol BS8 1TD, U.K.*
[2] *Division of Biotechnology, PHLS CAMR, Salisbury SP4 0JG, U.K.*

[Plate 1]

SUMMARY

Analysis of the mechanism and structure of lactate dehydrogenases is summarized in a map of the catalytic pathway. Chemical probes, single tryptophan residues inserted at specific sites and a crystal structure reveal slow movements of the protein framework that discriminate between closely related small substrates. Only small and correctly charged substrates allow the protein to engulf the substrate in an internal vacuole that is isolated from solvent protons, in which water is frozen and hydride transfer is rapid. The closed vacuole is very sensitive to the size and charge of the substrate and provides discrimination between small substrates that otherwise have too few functional groups to be distinguished at a solvated protein surface. This model was tested against its ability to successfully predict the design and synthesis of new enzymes such as L-hydroxyisocaproate dehydrogenase and fully active malate dehydrogenase. Solvent friction limits the rate of forming the vacuole and thus the maximum rate of catalysis.

INTRODUCTION

The rational design of new protein-based devices is possible because advanced computer-controlled chemical synthesis makes available DNA fragments of defined sequence, which can be biologically translated into a polypeptide chain of any chosen sequence of amino acids. Most biological properties of a protein depend upon the extended polypeptide chain folding into a unique three-dimensional structure. Synthetic polypeptide sequences do not usually fold fast enough to avoid tangled precipitates or survive attack by cellular degradation systems and thus protein engineers currently making new proteins avoid the kinetic folding problem by only altering small parts of otherwise large, stable and rapidly self-folding protein frameworks. This we call protein redesign as opposed to *de novo* protein design. Redesign requires a correct analysis of the natural protein design, but difficulties arise in distinguishing features important for the accepted properties (as an enzyme, hormone, etc.) from those important for more cryptic, often unsuspected biological functions. One way to avoid such red herrings is to test analytical conclusions against their success in predicting the synthesis of structures with predetermined functions. This paper describes this approach applied to the enzyme NAD[+]: L-lactate oxidoreductase.

ANALYSIS OF THE NATURAL ENZYME

The relation of function to the lactate dehydrogenase framework is summarized by a map of the catalytic pathway (figure 1). The established chemical mechanism is direct stereospecific transfer of a hydride ion from the C_4-position of the dihydronicotinamide ring to directly reduce a carbonyl (Westheimer 1987). The free protein obligatorily binds each coenzyme before the small substrate at a histidine–aspartate couple of defined protonation state. The diagram shows the rapid ($> 1000 \text{ s}^{-1}$) reversible two-electron equilibrium between a protonated histidine–aspartate couple through the keto—hydroxy-acid bond to the nicotinamide ring. This map explains many experimental observations, namely the neutral dihydropyridine ring of NADH binds much tighter than the positively charged ring of NAD[+]; ternary complexes of NADH and a ketoacid are compulsorily protonated and substrate can force a proton onto the histidine–aspartate couple against a 10^4 unfavourable bulk proton concentration (Holbrook 1973); dihydropyridine adducts formed from NAD[+] and sulphite, cyanide or enolpyruvate are stabilized by up to 10^5 compared with the stability free in solution (Parker & Holbrook 1977; Lodola *et al.* 1978); stabilization 'on-enzyme' of NADH and pyruvate by 10^3 over the free solution equilibrium (Holbrook & Gutfreund 1973; Clarke *et al.* 1988). These perturbed 'on-enzyme' equilibria reflect a very hydrophobic environment around the pyridine ring which, in an environment shielded from solvent, selects for neutral as opposed to charged adducts (Parker & Holbrook 1977). The tests included: making the pyridine ring environment more hydrophilic (Ile 250 → Asn) – this equalizes affinity for NAD[+] and NADH (Wigley *et al.* 1987a). Replacing Asp 168 by Ala or Asn selectively destabilizes the protonated His 195/NADH carbonyl groundstate compared to the neutral His 195/NAD[+]/carbinol groundstate (Clarke *et al.* 1987a) and pre-resonance laser Raman spectra show the carbonyl to then have only 20 % single bond character compared with 40 % single bond in the wild-type ground state (Deng *et al.*

Phil. Trans. R. Soc. Lond. B (1991) **332**, 177–184
Printed in Great Britain

[71]

177

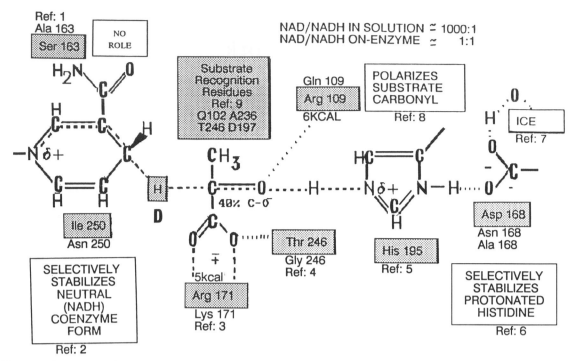

Figure 1. A map showing the roles of amino acids around the extended lactate dehydrogenase catalytic pathway of *B. stearotherophilus* lactate dehydrogenase. Original results are for Ref. 1: Wigley *et al.* 1987*b*. Ref. 2: Wigley *et al.* 1987*a*. Ref. 3: Hart *et al.* 1987*a*, *b*; Wigley *et al.* 1987*c*. Ref. 4: Clarke *et al.* 1987*a*, Bur *et al.* 1989. Ref. 5: Holbrook 1973; Holbrook & Ingram 1973; Lodola *et al.* 1978. Ref. 6: Clarke *et al.* 1988. Ref. 7: this paper. Ref. 8: Clarke *et al.* 1986. Ref. 9: Clarke *et al.* 1987*a*; Wilks *et al.* 1988, 1990.

1990). Arginine 109 (on the underside of the mobile surface loop) hardly stabilizes the stretched carbonyl in the groundstate, but compared with a glutamine-109 mutant, it decreases the transition state barrier by 4–6 kcal mol^{-1} (Clarke *et al.* 1986). Pathway residues are conserved in all LDH sequences but some, such as Ser 163, may be changed (e.g. to Ala 163) without measurable effect. Residues 109 and 102, which are important for catalysis and for substrate recognition, are on the underside of the mobile loop (98–110) and only take up the position shown after the substrate-induced conformation change.

A SUBSTRATE-INDUCED PROTEIN SHAPE CHANGE LIMITS CATALYSIS

To move a substrate into an internal, proton- and solvent-isolated vacuole requires the protein change between at least two shapes: one where the internal pocket is open to the solvent so the ionic substrate can enter and another closed state which has established the catalytic pathway. Kinetic experiments similar to that shown in figure 2*d*, plate 1 for pig M4 LDH show: (i) the maximum rate of conversion of pyruvate to lactate was uninfluenced by changing the hydrogen

DESCRIPTION OF PLATE 1

Figure 2. (*a–g*) Crystallographic and kinetic description of the substrate induced conformation change. (*a*) The packing of molecules of pig M$_4$ lactate dehydrogenase in crystal of the enzyme with NADH and oxamate from which half the oxamate has been washed out. Because of the 26° tilt between adjacent molecules in the crystal lattice, the mobile specificity loop (coloured white) of one of the two subunits in the asymmetric unit (coloured orange in (*a*)) is much closer to the next tetramer than that of the other subunit (red in (*a*)). In the orange subunit the specificity loop (picked out in grey) cannot open because it contacts the next tetramer in the lattice: this active site is still ternary i.e. contains both NADH and oxamate (*f*). In the red, unrestrained subunit the loop has partly opened and the active site contains no oxamate i.e. is a binary structure (*g*). (*b*) The Cα backbones of the binary (yellow) and ternary (deep blue) subunits superimposed. Red dots mark residues conserved in 21 sequences. The coenzyme and oxamate are coloured red. Where the subunits exactly match one grey trace is seen. (*c*) A schematic of a subunit of *B. stearothermophilus* lactate dehydrogenase complex with NADH, oxamate (picked out as a ball stick) and fructose-1,6-bisphosphate and oxamate with the secondary structures labelled. The bold trace is the specificity and catalysis loop (98–110), which opens and closes to release and accept substrate. (*d*) Stopped flow trace of the maximum rate of oxidation of NADH by pig M$_4$ enzyme and a saturating concentration of pyruvate, pH 7.2, 25 °C. This rate is 320 s^{-1} when NADD ([nicotinamide-4-^{2}H]NADH) is used instead of NADH. (*e*) The binding of saturated concentrations of a substrate–analogue, oxamate, is limited by the same conformation change, this time monitored by the change in tryptophan fluorescence. (*f*) The electron density around the ternary active site. Note the strong density for waters close to Glu 194, Arg 109, Asp 168 and Asn 140, which are absent or much weaker in the (*g*) electron density map for the binary complex.

Phil. Trans. R. Soc. Lond. B (1991)

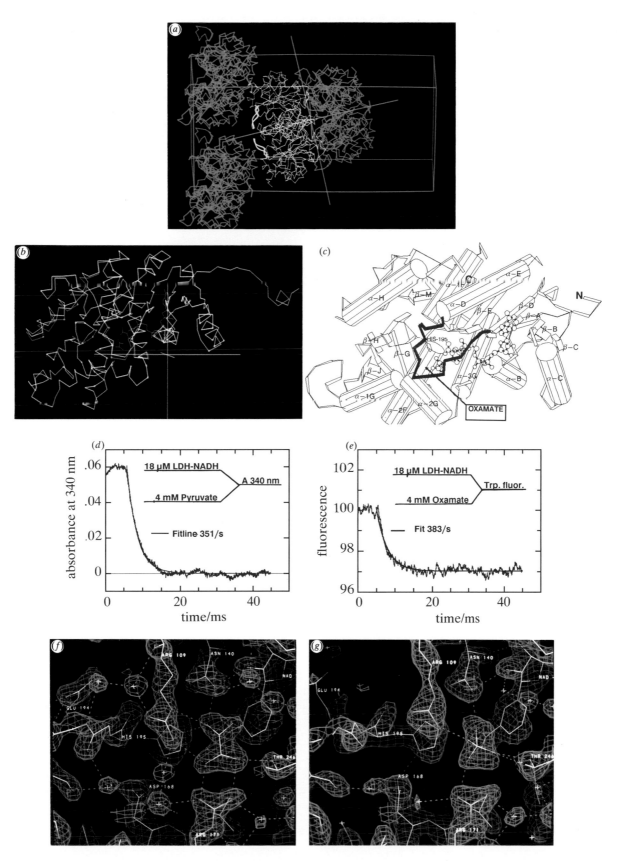

Figure 2. For description see opposite.

which is transferred to deuterium (whereas when C—H bond breaking is the slowest step, deuterium substitution slows rates by 3–7 fold, e.g. the slower mutant protein–substrate pairs of figure 4 *a*), and (ii) there was no rapid formation of the first mole of product (i.e. product release was not rate limiting). Thus it was deduced there must be an isomerization of the complex of protonated enzyme–NADH–pyruvate taking 1.5–5 ms before the chemical step in catalysis (Holbrook & Gutfreund, 1973).

A direct measure of shape change seen from a specific point is the change in the ionization of 3-nitro-tyrosine-237 on α-2G as substrate binding induces this helix to approach the acid of Asp 197 on the catalytic histidine loop (Parker *et al.* 1982; Clarke *et al.* 1985). Both these papers originally interpreted the acidification as being due to the approach of glutamate-107. However, Hart (1989) has shown that the α mutant with glutamine at 107 gives the full yellow colour change, whereas in a mutant with neutral asparagine at 197 the colour change is abolished. Thus acidification is due to the approach of Asp 197. The slower of two rates of yellow colour change matches k_{cat} over a wide range of temperatures and solvents.

Useful chemical modification is rare. A more general method is possible: all the natural tryptophan residues are removed from the protein before insertion of a single, fluorescent tryptophan at sites where motion is suspected (the motion of the specificity loop was directly monitored from a tryptophan introduced at the tip of the loop at position 106 and again showed that the maximum velocity of this catalyst was the same as the rate at which the tryptophan probe experienced an environment change in the loop closure process (Atkinson *et al.* 1987; Waldman *et al.* 1988)). The same techniques also enable intermediates during protein-folding to be characterized (Smith *et al.* 1990). In some cases changes in the environment of these probes are faster than the catalytic rate-limiting event and this helps to define the order of secondary structure movements.

DESCRIPTION OF STRUCTURE BEFORE AND AFTER THE SHAPE CHANGE

The fast spectroscopic probes defined the 3 ms shape change around Gly 106, Trp 248 and Cys 165 as being rate-limiting in catalysis; the initial change sensed by Tyr 237 ↔ Asp 197 is ten times faster. During the shape-change induced by binding of saturating concentrations of either the substrate, pyruvate or its inactive analogue, oxamate(H_2N—CO—COO⁻), the fluorescence of the tryptophan residues of the enzyme alters. The rate of this change matches the maximum rate of the enzyme reaction (figure 2 *d, e*). The structures before and after the shape change were measured. A single crystal of the pig M_4-enzyme–NADH–oxamate ternary complex was mounted in a microspectrofluorimeter and by observing the fluorescence of the dihydronictinamide ring it was possible to establish that half the oxamate could be washed out by using a buffer containing NADH and ammonium

sulphate without disordering the crystal. To obtain the full fluorescence of the pig M_4 LDH and NADH required further washing and the crystal normally cracked. The structures of both the full ternary complex (all subunits containing NADH and oxamate) and of the half binary–half ternary complex were determined by X-ray crystallography to 2.0 Å† (R-factor 0.256) and 2.25 Å (R-factor 0.253) resolution, respectively.

The crystal packing (figure 2 *a*) immediately revealed different contacts for two out of the four catalytic vacuoles. Two of the subunits were unrestrained to shape change on oxamate removal and their structures in the ternary (figure 2 *f*) and binary (figure 2 *g*) states yield a very accurate description of the shape change which accompanies the rate-limiting conversion of binary and ternary complex. The coordinates of the backbones of the two subunits are shown in figure 2 *b* superimposed. Figure 2 *c* enables the secondary structures which move to be named. Where a shape change has taken place the yellow backbone trace of the binary structure separates from the blue trace of the ternary state.

A sequence of conformational changes starting from the charge balancing when the negatively charged substrate enters the active site can be followed outwards to the surface movements of helices α2G, αD and αH. Tracing the pathway is helped by experimental kinetic information where it is available. At room temperature relating the changes to a substrate-induced trigger is reasonable, but at low temperature and in dimethyl-sulphoxide it is possible to induce a structure where the first turn-over of the enzyme is 10-times faster than the normal, conformationally restrained k_{cat} (Clarke *et al.* 1985; Atkinson *et al.* 1987). Conserved amino acid residues (picked out as red dots in figure 2 *b*), which are at the contact points between the secondary structures that move are listed in table 2. The fastest event is the approach of Asp 197 to Tyr 237 (α2G). The clenching of the specificity loop (98–110) against α2G is slower when seen from a tryptophan inserted at position 106 (Waldman *et al.* 1988). Substrate-binding bridges His 195 and Thr 246, draws in Asp 197 and enables withdrawal of Glu 194 from a hydrophobic pocket formed by Leu 322 and Ile 325 on αH. Movement enables Ser 198 to drag (via an H-bond to Ser 318 on αH) the entire C-terminal helix 1.5 Å in towards the active site. The net result of the changes is a tighter interlocking of the external elements of the active site (the βG/βH turn, the α1G/α2G helix and the active site loop) such that the catalytic groups His 195 and Arg 109 are shielded from the effects of bulk solvent. Many conserved residues occur at the contact points where secondary structures move in the substrate-induced change and in the catalytic centre. Helix αC, which appears remote, donates His 68 to the active site of the Q-axis related subunit.

In the ternary complex (figure 2 *f*) there are many ordered waters in the catalytic vacuole which form extra stabilizing bridges with the inside surface of the specificity loop and with some of the residues whose

† 1 Å = 10^{-10} m = 10^{-1} nm.

Table 1. *A very active malate dehydrogenase from a lactate dehydrogenase framework* (*from Wilks* et al. *1988*)

enzyme	substrate	$\dfrac{k_{cat}}{s^{-1}}$	$\dfrac{K_M}{mM}$	$\dfrac{k_{NADH}}{k_{NADD(\pm 0.1)}}$	$\dfrac{t_{0.5}90\,°C}{m}$	$\dfrac{K_M}{K_M(+FBP)}$	$\dfrac{k_{cat}/K_M(OAA)}{k_{cat}/K_M(PYR)}$
Q102 WT	pyruvate	250	0.06	1.1	7	50	
	oxaloacetate	6	1.5	2.8	7	n.d.	0.95×10^{-3}
R102	pyruvate	0.9	1.8	2.6	6.5	2.8	
	oxaloacetate	250	0.06	1.1	6.5	33	8.4×10^{3}

position alters in the shape change. Fewer low B-value waters are seen in the more solvent-exposed binary complex (figure 2g). The strongly ordered H-bonds from these frozen waters to elements of the coenzyme and bound substrate–analogue reduce the number of conformers available along the catalytic pathway to facilitate fast 2-electron transfer over 20 Å (figure 1).

TEST OF ANALYSIS: SYNTHESIS OF A MALATE DEHYDROGENASE ON THE LACTATE DEHYDROGENASE FRAMEWORK

The analysis of specificity suggests only small and singly negatively charged ketoacids enable the specificity loop to close correctly onto α-2G and enable the catalytic Arg 109 to approach the substrate carbonyl

and stabilize the transition state by 4–6 kcal mol^{-1}. This analysis was used to suggest designs of an enzyme that would recognize and transform a substrate with two negative charges: that is oxaloacetate instead of pyruvate:

$$
\begin{array}{cc}
 & COO^- \\
 & | \\
CH_3 & CH_2 \\
| & | \\
C{=}O & C{=}O \\
| & | \\
COO^- & COO^- \\
\text{pyruvate} & \text{oxaloacetate}
\end{array}
$$

A fixed-volume vacuole which can only accommodate balanced charges suggests three ways of making the

Table 2. *Conserved amino acid residues in the secondary structure element contacts which move in the binary* (*E-NADH*) *to ternary* (*E-NADH-oxamate*) *conformation change*

(Residues are identical in at least 21 aligned primary structures (J. J. Holbrook, unpublished alignment). This table and figure 2b, c enable the bonds that transmit conformation information from the surface to the active centre and vice versa to be identified. The coenzyme does not move in this shape change.)

secondary structures	residues	Cα separation binary Å	ternary	bond type
βG–α3G	His 195–Thr 246	14.3	13.9	H-bond via oxamate
βH–βG	Ser 198–Glu 194	6.9	5.9	H-bond
	Asp 197–Glu 194	7.9	7.4	H-bond via water
αH–βH	Ser 318–Ser 198	7.0	5.3	H-bond
	Glu 311–Ser 204	9.8	8.5	H-bond
αH–βD/αD (loop)	Leu 322–Leu 110	9.6	9.1	Hydrophobic
βG–βD/αD	Glu 194–Arg 109	11.5	10.5	H-bond via 2 waters
α1G/α2G–βD/αD	Tyr 237–Arg 109	10.7	10.4	H-bond via water
	Val 232–Arg 109	15.6	12.5	H-bond via 2 waters
βG–α1G/α2G	Asp 197–Tyr 237	10.3	10.3	H-bond
βE/α1F–βG	Asp 143–His 195	6.9	7.0	H-bond
βE/α1F–coenzyme	Asn 140–ribose C_2	4.7	4.6	H-bond
	Val 138–amide C_7	5.7	5.5	H-bond
	Val 138–pyridine N_1	5.8	6.0	Hydrophobic
α2F–coenzyme	Ser 163–amide	6.6	6.5	H-bond via water
	Gly 164–amide	7.0	6.8	H-bond via water
βG–α2F	His 195–Asp 168	7.1	6.8	H-bond
βL–αH	Leu/Val 285–Ala 319	6.0	6.3	Hydrophobic
	Ile/Leu 287–Ala 319	7.3	7.1	Hydrophobic
coenzyme–βD/αD	ribose–Ala 100	6.4	6.2	H-bond
	ribose–Gln 102	6.8	6.5	H-bond via water
coenzyme–βA/αB	pyridine–Val 32	6.1	6.4	Hydrophobic
coenzyme–α3G	pyridine–Ile 250	7.3	7.3	Hydrophobic
α3G–αC (Q-axis)	Trp 248–Glu 62	9.3	8.2	H-bond
α2F–αC (Q-axis)	Arg 171–His 68	5.2	5.1	H-bond

new activity on this framework (Wilks *et al.* 1988). Increasing vacuole size by changing the bulky Thr 246 to Gly 246 resulted in a greater selectivity for the new target substrate albeit at the expense of a much reduced k_{cat} (loss of H-bond linking His 195 to α-3G). The extra negative charge of the new target substrate was partially balanced by removing non-catalytic negative charge from the periphery of the vacuole (either Asp 197 → Asn or Glu 107 → Gln). Both these mutations shifted selectivity in favour of the new target substrate. Introducing a positively charged amino acid (Gln 102 changed to Arg 102) on the inside surface of the vacuole adjacent to the modelled position of the new target carboxylate had a large effect (table 1). In this construction the old, natural substrate pyruvate cannot provide a counter-ion for the new arginine 102: it is discriminated against by about 950-fold; whereas the new arginine can be solvated by the negatively charged oxaloacetate and is favourably selected by 8400-fold. The net result is a switch against pyruvate and in favour of oxaloacetate by nearly 10^7-fold. Like the wild-type enzyme and substrate, the new framework is limited in rate by a shape change and not by chemistry ($k_H/k_D = 1.1$). In contrast to many previous enzyme redesigns, this new construct is fast (250 s^{-1}) and has a low K_M (60 μM) and unlike natural malate dehydrogenases, the new design is allosterically activated 33-fold by fructose-1,6-bisphosphate.

SYNTHESIS OF A BROAD SPECIFICITY HYDROXYACID DEHYDROGENASE

One use for enzymes is the synthesis of single compound drugs and herbicides. Such drugs make toxicology testing simpler and (since the tragedy of an apparently safe drug for controlling morning sickness being found to contain an enantiomer that impaired foetal development) drug houses are increasingly only bringing forward single enantiomers for approval for human use (de Camp 1989). Natural lactate dehydrogenase makes 100% enantiomeric excess *s*-lactates, but is of limited use in chiral synthesis since most drug intermediates have large side chains that disrupt the fixed volume vacuole. The enzyme is almost unusable with branched side chains (figure 3*b*, solid line). A cross-section of the specificity loop and α2G (figure 2*a*) suggests a wide range of larger pyruvates might be converted to chiral lactates if the 'JAW' region could be made sufficiently hydrophobic and flexible to accommodate a range of branched chain paraffins. In the lower region (α2G) alanines 235–236 were replaced by glycines to both increase the volume for large side chains and to provide conformational flexibility by destabilizing the helix. As can be seen from figure 3*b* this change alone gave only two- to five-fold improvement in k_{cat} for the largest substrates. Making the hydrophilic inside surface of the upper 'JAW' more hydrophobic (Gln–Lys 102–3 to Met–Val) and the introduction of a serine instead of proline-105 to add flexibility also provided only marginal improvements in k_{cat} (for large-scale production of chiral products improved k_{cat} is more important than a low K_M since the high substrate concentration is normally used to

aid product isolation). However, when both sets of mutations are made a reasonable α-hydroxyisocaproate dehydrogenase results with a k_{cat} of 18.5 s^{-1} (compared with wild-type k_{cat} of 250 s^{-1} for pyruvate). The improvement was obtained without trial and error. It is unlikely random mutagenesis and selection would produce mutations in two coupled surfaces at once.

A MUTANT WITH ALTERED CONFORMATION CHANGE

In making these new constructions on the lactate dehydrogenase framework the mobile loop was altered and structures with changed rate of the conformation change may have resulted. However, the slower mutants are limited in their maximum rate by the rate of the chemical reduction and show a kinetic deuterium isotope ratio of 2–4 (figure 4*a*). The only experimental information then available is that the shape change (k_{conf}) is faster than the rate of the chemical change (k_{chem}). Most of the results fall in an envelope which increases to 250 s^{-1}. However, for one construction (QKP 102–5 → MVS) the k_{cat} with pyruvate and α-ketobutyrate are slowed to 100 s^{-1}, but without C—H bond breaking becoming rate-limiting (primary deuterium isotope ratio is 1.2). This mutation, which slows the rate of the conformation change involves the inside surface of the mobile specificity loop. The slowing mutation is reversed when the surface against which it closes is made more open when the AA → GG mutation is also recruited. This again suggests the final closure of the specificity loop against α-2G is important in the slowest shape change (250 s^{-1}) and lags in time the fast approach of Asp 197 to Tyr 237 on α2G (estimated at 3000 s^{-1} at 25°; Atkinson *et al.* 1987).

PENALTIES OF A VACUOLE MECHANISM. THE RATE OF THE SUBSTRATE-INDUCED SHAPE CHANGE IS SLOWED BY SOLVENT FRICTION

The method used to prepare the half binary/half ternary crystal shows that if the specificity loop (98–110) movement is prevented then the reordering of α2G and αD and αH are also prevented. Viscous solutions slow the k_{cat} of many enzymes whose rate is conformationally limited and Saburova *et al.* (1988) observed 44% glycerol reduces k_{cat} for pig M_4 lactate dehydrogenase from 350 s^{-1} to 35 s^{-1}. Elements which might transmit the surface frictional drag to the active centre are the conserved bonds between αH and βH (Ser 318–Ser 198, Glu 311–Ser 204) and between αH and αD (Leu 322–Leu 110). Thus in the malate dehydrogenase of *Thermus flavus* which has the same framework as lactate dehydrogenase a spontaneous mutant (Thr 190 to Ile) arose (breaks an H-bond bridge to Thr 315 of α-H) with the k_{cat} increased from 460 s^{-1} to 1200 s^{-1} (Nishiyama *et al.* 1986). An H-bond is also possible between the homologous pair of conserved residues in lactate dehydrogenase (Ser 198 ...Ser 318). This fortuitous mutant suggests it may be unsafe to extend Darwin's ideas on survival of the

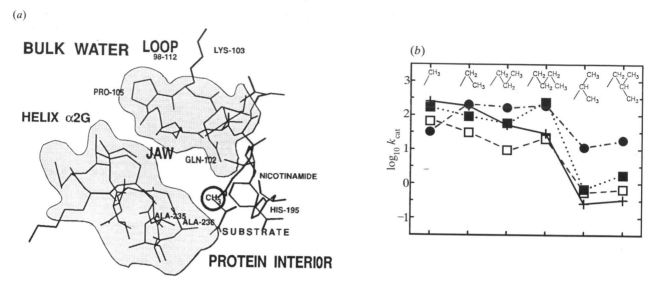

Figure 3. The lactate dehydrogenase catalytic vacuole used to design a broad specificity hydroxyacid dehydrogenase. (*a*) A cross section through the catalytic vacuole showing the two moving surfaces (loop and helixα2G) which seal the vacuole from the solvent. (*b*) A plot of $\log_{10}(k_{cat})$ for a series pyruvates ($R.CO.COO^-$) for the wild-type enzyme (—+—), a mutant with $^{235-6}AA \rightarrow GG$ (lower jaw --■--), a mutant with $^{102-105}QKP \rightarrow MVS$ (upper jaw ...□...), and all five mutations at once (-----●-----). Only when both surfaces are changed together is a broad specificity enzyme formed. Replotted from Wilks *et al.* (1990). The fixed size vacuole of the wild-type will not accommodate side chains larger than $—CH_2—CH_3$.

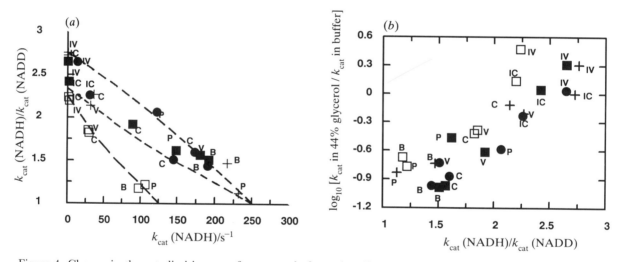

Figure 4. Change in the rate-limiting step from vacuole formation (fast enzyme–substrate pairs) to bond breaking (slow pairs). (*a*) The primary deuterium kinetic isotope ratio as a function of the k_{cat} with NADH for the four enzymes of figure 3*b*: wild-type (+), $^{102-105}MVS$ (□), $^{235-236}GG$ (■) and all five changes together (●) with P, pyruvate; B, α-ketobutyrate; V, α-ketovalerate; C; α-ketocaproate; IV, α-ketoisovalerate; IC, α-ketoisocaproate.

$$\text{NADH} + \text{pyruvate} + \text{H}^+ \rightarrow \underset{\substack{\text{fast} \\ >C=O}}{E—BH^+} \overset{\ulcorner\text{NADH}}{\rightarrow} \underset{\substack{k_{conf} \\ >C=O}}{\overset{*}{E}—BH^+} \overset{\ulcorner\text{NADH}}{\rightarrow} \underset{\substack{k_{chem} \\ —HC—OH}}{E—B:} \overset{\ulcorner\text{NAD}^+}{\rightarrow} E + \text{NAD}^+ + \text{lactate}.$$

The lines are drawn according to a reverse reaction mechanism (Holbrook *et al.* 1975), which has two steps after the rapid bimolecular binding of first coenzyme and then substrate: the isomerization of the initially formed ternary complex (k_{conf}) followed by the chemical step (k_{chem}) in which hydride is transferred and NAD^+ and lactate are formed on enzyme and then rapidly released into solution. The envelope lines (------) model k_{chem} (NADH)/k_{chem} (NADD) = 2.4–2.8 and $k_{conf} = 250$ s^{-1} and enclose results for the wild type, $^{235-236}GG$, and the mutant enzyme with all five changes. The line (— — —) models the conformational change mutant $^{102-105}MVS$ with isotope ratio = 2.3 and $k_{conf} = 125$ s^{-1}. (*b*) The slowing of maximum velocity by 44% (w/v) glycerol. The decadic logarithm of the ratio of the maximum rate of catalysis of the same constructions and substrates were measured in buffer and in 44% glycerol. Fast enzyme–substrate pairs where conformation change rate-limits are markedly slowed by glycerol. Slow pairs, where chemistry limits show a slight rate increase. The general trend is visible from 24 enzyme–substrate pairs.

fittest by adaptation (Darwin 1858) to include perfect adaption of individual biomolecules. Lactate dehydrogenase shows improvement of a natural enzyme for an unnatural substrate (see also Clarke *et al.* 1987*b*), but we may now seek and find modest improvement against natural substrates (e.g. the Asp 53 → Ala enzyme uses both NAD^+ and $NADP^+$ (Feeney *et al.* 1989) for cheap NADPH coenzyme regeneration in enzyme reactors).

The effect of bulk viscosity is consistently seen in all the 24 enzyme and substrate pairs of figure 3*b*. Where the framework–substrate pairing gives slow catalysis limited by bond breaking ($k_{H/D} > 1$ in figure 4*a*) the effect of viscosity is small. However, for fast enzyme and substrate pairs limited by the substrate-induced shape change the effect of glycerol is to slow k_{cat} by 5–10-fold. The slowing in glycerol is inversely correlated with the appearance of the primary deuterium isotope effect (figure 4*b*) and strongly supports the concept that, in the natural enzyme, part of the free energy decrease on substrate- or substrate analogue-binding is dissipated in driving the movement of surface peptide structures against friction of the solvent. This system is one of few where description of the coupling of the energy of substrate-binding to the ability of a protein to do work is possible in atomic detail.

We thank The SERC, Genzyme (U.K.) Ltd, Porton Industries, NATO (travel) and SmithKline-Beecham for support.

REFERENCES

Atkinson, T., Barstow, D. A., Chia, W. N., Clarke, A. R., Hart, K. W., Waldman, A. D. B., Wigley, D. B., Wilks, H. M. & Holbrook, J. J. 1987 Mapping motion in large proteins by single tryptophan probes inserted by site-directed mutagenesis: lactate dehydrogenase. *Biochem. Soc. Trans.* **15**, 991–93.

Bur, D., Clarke, A. R., Friesen, J. D., Gold, M., Hart, K. W., Holbrook, J. J., Jones, J. B., Luyten, M. A. & Wilks, H. M. 1989 On the Effect on Specificity of Thr246 → Gly Mutation in L-lactate dehydrogenase of *Bacillus stearothermophilus*. *Biochem. biophys. Res. Commun.* **161**, 59–63.

Clarke, A. R., Wilks, H. M., Barstow, D. A., Atkinson, T., Chia, W. N. & Holbrook, J. J. 1988 An investigation of the contribution made by the carboxylate group of an active site histidine–aspartate couple to binding and catalysis in lactate dehydrogenase. *Biochemistry* **27**, 1617–1622.

Clarke, A. R., Smith, C. J., Hart, K. W., Birktoft, J. J., Banaszak, L. J., Wilks, H. M., Barstow, D. A., Atkinson, T., Lee, T. V., Chia, W. N. & Holbrook, J. J. 1987*a* Rational construction of a 2-hydroxyacid dehydrogenase with new substrate specificity. *Biochem. biophys. Res. Commun.* **148**, 15–23.

Clarke, A. R., Wigley, D. B., Barstow, D. A., Chia, W. N., Atkinson, T. & Holbrook, J. J. 1987*b* A single amino acid substitution deregulates a bacterial lactate dehydrogenase and stabilizes its tetrameric structure. *Biochim. biophys. Acta* **913**, 72–80.

Clarke, A. R., Wigley, D. B., Chia, W. N., Barstow, D., Atkinson, T. & Holbrook, J. J. 1986 Site-directed mutagenesis reveals the role of a mobile arginine residue in lactate dehydrogenase catalysis. *Nature, Lond.* **324**, 699–702.

Clarke, A. R., Waldman, A. D. B., Hart, K. & Holbrook, J. J. 1985 The rates of defined changes in protein structure during the catalytic cycle of lactate dehydrogenase. *Biochim. biophys. Acta* **829**, 397–407.

Darwin, C. R. 1859 *The origin of species* or *the preservation of favoured races in the struggle for life*. London: John Murray.

De Camp, W. H. 1989 The Food and Drug Administration's Perspective on the development of stereoisomers. *Chirality* **1**, 2–6.

Deng, H., Zheng, J., Burgner, J., Clarke, A. R., Holbrook, J. J. & Callender, R. H. 1990 A Raman spectroscopic study of pyruvate bound to lactate dehydrogenase and its R109Q and D168N Mutants. *Biophys. J.* **57**, 41.

Feeney, R., Clarke, A. R. & Holbrook, J. J. 1989 A single amino acid substitution in lactate dehydrogenases improves the catalytic efficiency with an alternative coenzyme. *Biochem. biophys. Res. Commun.* **166**, 667–672.

Hart, K. W. 1989 An investigation of the molecular basis of substrate specificity in lactate dehydrogenase. Ph. D. thesis, University of Bristol.

Hart, K. W., Clarke, A. R., Wigley, D. B., Chia, W. N., Barstow, D. A., Atkinson, T. & Holbrook, J. J. 1987*a*. The importance of arginine-171 in substrate binding by *Bacillus stearothermophilus* lactate dehydrogenase. *Biochem. biophys. Res. Commun.* **146**, 346–53.

Hart, K. W., Clarke, A. R., Wigley, D. B., Waldman, A. D. B., Chia, W. N., Barstow, D. A., Atkinson, T., Jones, J. B. & Holbrook, J. J. 1987*b*. A strong carboxylate-arginine interaction is important in substrate orientation and recognition in lactate dehydrogenase. *Biochim. biophys. Acta* **914**, 294–298.

Holbrook, J. J., Liljas, A., Steindel, S. J. & Rossmann, M. G. 1975 Lactate dehydrogenase. *Enzymes* **11a**, 191–293.

Holbrook, J. J. 1973 Direct measurement of proton binding to the active ternary complex of pig heart lactate dehydrogenase. *Biochem. J.* **133**, 847–849.

Holbrook, J. J. & Gutfreund, H. 1973 Approaches to the study of enzyme mechanisms: lactate dehydrogenase. *FEBS Lett.* **31**, 157–169.

Holbrook, J. J. & Ingram, V. A. 1973 Ionic properties of the essential histidine in pig heart lactate dehydrogenase. *Biochem. J.* **131**, 729–738.

Nishiyama, M., Matsubara, N., Yamamoto, K., Iijima, S., Uozumi, T. & Beppu, T. 1986 Nucleotide sequence of the malate dehydrogenase gene of *Thermus flavus* and its mutation directing an increase in enzyme activity. *J. biol. Chem.* **261**, 14178–14183.

Lodola, A., Parker, D. M., Jeck, R. & Holbrook, J. J. 1978 Malate dehydrogenase of the cytosol. Ionizations of the enzyme-reduced-coenzyme complex and a comparison with lactate dehydrogenase. *Biochem. J.* **173**, 597–605.

Parker, D. M., Jeckel, D. & Holbrook, J. J. 1982 Slow structural changes shown by the 3-nitrotyrosine237 residue in pig heart [Tyr(3NO$_2$)^{237}] lactate dehydrogenase. *Biochem. J.* **201**, 465–471.

Parker, D. M. & Holbrook, J. J. 1977 An oil-water: histidine mechanism for the activation of coenzyme in the α-hydroxyacid dehydrogenases. In *Pyridine nucleotide-dependent dehydrogenases* (ed. H. Sund), pp. 485–495. New York and Berlin: Walter de Gruyter.

Saburova, E. K., Kamenchuk, O. I. & Demchenko, A. P. 1988 Kinetics of the reaction catalyzed by lactate dehydrogenase: a dynamic aspect. *Molec. Biol. Moscow* **22**, 718–725.

Smith, C. J., Clarke, A. R., Chia, W. N., Irons, L. I., Atkinson, T. & Holbrook, J. J. 1990 Detection and characterization of intermediates in the folding large proteins by the use of genetically inserted tryptophan probes. *Biochemistry* **30**, 1028–1036.

Phil. Trans. R. Soc. Lond. B (1991)

Waldman, A. D. B., Hart, K. W., Clarke, A. R., Wigley, D. B., Barstow, D. A., Atkinson, T., Chia, W. N. & Holbrook, J. J. 1988 A genetically engineered single tryptophan identifies the movement of a peptide domain of lactate dehydrogenase as the event which limits maximum enzyme velocity. *Biochem. biophys. Res. Commun.* **150**, 752–759.

Westheimer, F. H. 1987 Mechanism of action of the pyridine nucleotides. In *Coenzymes and cofactors, volume IIA – pyridine nucleotide coenzymes* (ed. D. Dolphin, O. Avramoviĉ & R. Poulson), pp. 253–322. New York: John Wiley.

Wigley, D. B., Clarke, A. R., Dunn, C. R., Barstow, D. A., Atkinson, T., Chia, W. N., Muirhead, H. & Holbrook, J. J. 1987 *a* The engineering of a more thermally stable lactate dehydrogenase by reduction of the area of a water-accessible hydrophobic surface. *Biochim. biophys. Acta* **916**, 145–148.

Wigley, D. B., Clarke, A. R. & Holbrook, J. J. 1987 *b* Hydrogen bonding of the carboxyamide of NADH is not important for catalysis in lactate dehydrogenase. *Protein Eng.* **1**, 260.

Wigley, D. B., Lyall, A., Hart, K. W. & Holbrook, J. J. 1987 *c* The greater strength of arginine : carboxylate over lysine : carboxylate ion pairs. Implications for the design of novel enzymes and drugs. *Biochem. biophys. Res. Commun.* **149**, 927–929.

Wilks, H. M., Halsall, D. J., Atkinson, T., Chia, W. N.,

Clarke, A. R. & Holbrook, J. J. 1990 Designs for a broad substrate specificity α-ketoacid dehydrogenase. *Biochemistry* **29**, 8587–8591.

Wilks, H. M., Hart, K. W., Feeney, R., Dunn, C. R., Muirhead, H., Chia, W. N., Barstow, D. A., Atkinson, T., Clarke, A. R. & Holbrook, J. J. 1988 A specific and highly active malate dehydrogenase by redesign of a lactate dehydrogenase framework. *Science, Wash.* **242**, 1541–1544.

Discussion

A. J. Crumpton (*Fairways, Middle Street, East Harptree, Avon, U.K.*). Professor Blundell alluded to the fact that protein cores were evolutionarily more conserved than the protein surface.

In the final discussion I mentioned that when an active centre was first discovered upon an enzyme it had been necessary to use a non-polar solvent (acetone, free of both water and carbon dioxide, which I had prepared) for the ultraviolet spectroscopy of ribonuclease A.

I asked if any enzymologists present were using non-polar solvents to study the hydrophobic cores of proteins. A few did then report such work.

Plate 1 was printed by George Over Ltd., Rugby, U.K.